HIP CAPITALISM

Volume 81, Sage Library of Social Research

Sage Library of Social Research

1. Caplovitz **The Merchants of Harlem**
2. Rosenau **International Studies & the Social Sciences**
3. Ashford **Ideology & Participation**
4. McGowan/Shapiro **The Comparative Study of Foreign Policy**
5. Male **The Struggle for Power**
6. Tanter **Modelling & Managing International Conflicts**
7. Catanese **Planners & Local Politics**
8. Prescott **Economic Aspects of Public Housing**
9. Parkinson **Latin America, the Cold War, & the World Powers, 1945-1973**
10. Smith **Ad Hoc Governments**
11. Gallimore et al **Culture, Behavior & Education**
12. Hallman **Neighborhood Government in a Metropolitan Setting**
13. Gelles **The Violent Home**
14. Weaver **Conflict & Control in Health Care Administration**
15. Schweigler **National Consciousness in Divided Germany**
16. Carey **Sociology & Public Affairs**
17. Lehman **Coordinating Health Care**
18. Bell/Price **The First Term**
19. Alderfer/Brown **Learning from Changing**
20. Wells/Marwell **Self-Esteem**
21. Robins **Political Institutionalization & the Integration of Elites**
22. Schonfeld **Obedience & Revolt**
23. McCready/Greeley **The Ultimate Values of the American Population**
24. Nye **Role Structure & Analysis of the Family**
25. Wehr/Washburn **Peace & World Order Systems**
26. Stewart **Children in Distress**
27. Dedring **Recent Advances in Peace & Conflict Research**
28. Czudnowski **Comparing Political Behavior**
29. Douglas **Investigative Social Research**
30. Stohl **War & Domestic Political Violence**
31. Williamson **Sons or Daughters**
32. Levi **Law & Politics in the International Society**
33. Altheide **Creating Reality**
34. Lerner **The Politics of Decision-Making**
35. Converse **The Dynamics of Party Support**
36. Newman/Price **Jails & Drug Treatment**
37. Abercrombie **The Military Chaplain**
38. Gottdiener **Planned Sprawl**
39. Lineberry **Equality & Urban Policy**
40. Morgan **Deterrence**
41. Lefebvre **The Structure of Awareness**
42. Fontana **The Last Frontier**
43. Kemper **Migration & Adaptation**
44. Caplovitz/Sherrow **The Religious Drop-Outs**
45. Nagel/Neef **The Legal Process: Modeling the System**
46. Bucher/Stelling **Becoming Professional**
47. Hiniker **Revolutionary Ideology & Chinese Reality**
48. Herman **Jewish Identity**
49. Marsh **Protest & Political Consciousness**
50. LaRossa **Conflict & Power in Marriage**
51. Abrahamsson **Bureaucracy or Participation**
52. Parkinson **The Philosophy of International Relations**
53. Lerup **Building the Unfinished**
54. Smith **Churchill's German Army**
55. Corden **Planned Cities**
56. Hallman **Small & Large Together**
57. Inciardi et al **Historical Approaches to Crime**
58. Levitan/Alderman **Warriors at Work**
59. Zurcher **The Mutable Self**
60. Teune/Mlinar **The Developmental Logic of Social Systems**
61. Garson **Group Theories of Politics**
62. Medcalf **Law & Identity**
63. Danziger **Making Budgets**
64. Damrell **Search for Identity**
65. Stotland et al **Empathy, Fantasy & Helping**
66. Aronson **Money & Power**
67. Wice **Criminal Lawyers**
68. Hoole **Evaluation Research & Development Activities**
69. Singelmann **From Agriculture to Services**
70. Seward **The American Family**
71. McCleary **Dangerous Men**
72. Nagel/Neef **Policy Analysis: In Social Science Research**
73. Rejai/Phillips **Leaders of Revolution**
74. Inbar **Routine Decision-Making**
75. Galaskiewicz **Exchange Networks & Community Politics**
76. Alkin/Daillak/White **Using Evaluations**
77. Sensat **Habermas & Marxism**
78. Matthews **The Social World of Old Women**
79. Swanson/Cohen/Swanson **Small Towns & Small Towners**
80. Latour/Woolgar **Laboratory Life**
81. Krieger **Hip Capitalism**

HIP CAPITALISM

SUSAN KRIEGER

Foreword by James G. March

Volume 81
SAGE LIBRARY OF
SOCIAL RESEARCH

 **SAGE PUBLICATIONS**   Beverly Hills   London

Copyright © 1979 by Sage Publications, Inc.

All rights reserved. No part of this book may be reproduced or utilized in any form or by means, electronic or mechanical, including photocopying, recording, or by any information storage and retrieval system, without permission in writing from the publisher.

For information address:

SAGE Publications Inc.
275 South Beverly Drive
Beverly Hills, California 90212

SAGE Publications Ltd
28 Banner Street
London EC1Y 8QE England

Printed in the United States of America

Library of Congress Cataloging in Publication Data

Krieger, Susan.
   Hip capitalism.

   (Sage library of social research; v. 81)
   1. Radio stations—Social aspects—United States.
   2. Radio stations—United States—Employees.
   3. Rock music—United States.   4. Organizational behavior—
Case studies.   I. Title.

HE8698.K74   384.54'06'575   79-4554
ISBN 0-8039-1262-5
ISBN 0-8039-1263-3 pbk.

FIRST PRINTING

Contents

Foreword by James G. March	9
Acknowledgments	17
Introduction	19
Chronology	25
Characters	27

PART I: BEGINNINGS
Chapter

1	KMPX (March-June 1967)	31
2	KMPX (April-November 1967)	43
3	KMPX and KPPC (November 1967-March 1968)	57
4	The KMPX Strike (March-May 1968)	71
	March	71
	April	87
	May	96

PART II: LEGITIMACY
Chapter

5	KSAN (May-August 1968)	105
6	KSAN (September 1968-March 1969)	115
7	A Voice of the Revolution (April-August 1969)	129

PART III: PROFESSIONALISM
Chapter

8	A Voice of the Revolution (September-December 1969)	145
9	KSAN (January-May 1970)	165
10	KSAN (June-October 1970)	177
11	A Different Station (November 1970-May 1971)	193
12	KSAN (June-October 1971)	209

PART IV: RENEWAL
Chapter
13 An Incident (November 1971-January 1972) 225
14 KSAN (November 1971-April 1972) 239
15 KSAN (May-July 1972) 251
    The General Manageship 251
    Conditions 267
    The Staff 273

Conclusion 289

Sources 293

Appendix
  Cast of Characters 301
  Summary of Sales Growth 303
About the Author 304

For my father

FOREWORD

When William F. Whyte wrote *Street Corner Society*, he combined three things: a portrait of a decade that deeply affected the thought of a generation; a story of ordinary lives within a symbol of that decade, an urban gang; and an ethnographic method that illuminated the broader meaning of elementary human events. Susan Krieger has written a similar book about a different decade and a different symbol, the decade beginning in the mid-sixties and a rock radio station. She paints that history with the loving brush of one who knew the culture, was part of it, and wants to describe it in a way that neither destroys the uniqueness of the radio station, demeans the lives involved in it, nor compromises the standards of a professional social scientist. It is not an easy task.

The book is a short history and an austere memoir. It is careful and cautious, deliberately understated, an exhibition of reportorial discipline. Without artifice and without pseudonym, it tells a story; and it tells it well. Because I have considerable enthusiasm for the spirit and content of her book, I may, perhaps, exercise my right to accuse Professor Krieger of being unreasonably merciless in her resistance to saying what the book is about. At least I feel no hesitation in berating her, because I have been doing it for several years, without notable effect. The opportunity to write this foreword, however, gives me a new weapon — the chance to suggest what her book is about before she does.

This book is about organizations and how they confront their members with minor choices, elusive in significance and ambiguous in consequence, that collectively affect the direction of organizational drift and the development of personal meaning. It is about the containment and contagion of socially outrageous behaviors and beliefs, one of the most fundamental and pervasive of social processes. It is about an era, and no one who recalls the late 1960s will read it without nostalgia. It is about the writing of social science and the way it is related to the writing of budgets and poetry. It is about all of those things, but it talks of none of them.

The study is a history of an organization, of what happened to a radio station over a brief but compelling period. It was a normal, anarchic organization. It started as one radio station and ended as another one, leaving the first station still functioning. It did not have well-defined boundaries; people moved in and out. It is hard to be sure whom to treat as part of the organization and whom to treat as part of the environment. The organization had no clear goals and no clear plan. There were some notions of what good music was, what a proper disc jockey did, and what good advertising spots were like; but they did not provide a clear criterion for judging organizational performance, much less a criterion that was generally accepted. It was the kind of organization found commonly in the real world, less commonly in our theories.

The history is not an anonymous description of what happened at the radio station. It is an interweaving of accounts of what happened by people who were there. If a police file were written well, it would read like this, filtering the commentary through a screen that leaves almost everything there, but keeps the observer and the reader at a distance. Krieger narrates the stories of participants about each important event in cool counterpoint to the fervor they report. The stories are there with all of their inconsistencies, shadings, reconstructions, and self-acknowledgments. They honor the frayed collars of everyday life. Feelings, observations, inferences, and inventions are twisted together the way they usually are to provide a report that recognizes the complexities of meaning in a social situation. Through her skill in relating these accounts, including the accounts told by records, letters, ledgers, and memoranda, Professor Krieger exposes the texture of organizational life.

Students of organizations, social movements, and the history of the 1960s in the United States will find insights in the book. However, they will not find them labeled as theory or condensed into a set of testable propositions. Krieger has a different vision of the way social science and social history are written. Most contemporary work in social science presumes that the theoretical promises and implications of empirical work should be explicit. It is dogma for journal referees and dissertation committees. We are suspicious of hidden assumptions and scornful of "mere description." These beliefs lead sometimes to such a glorification of theory that empirical data become unrecognizable, and in some recent case studies of institutions it is almost impossible to see the events through the interpretation. But most good case studies in social science and history try to occupy a middle ground between history mangled by theoretical pretense and history swamped by uninterpreted data. Most of us try to be gentle with data and sensitive to their context, yet attempt to say something that is more general than their recitation. Krieger does not take that middle ground. Her book has a distinctive methodological stance. As rigorously as anyone I know, she avoids interpretation, generalization, and conceptualization.

The stance is not an accident, and it has not been adopted innocently.

It is a carefully considered and deliberate statement about empirical research in social science. Characteristically, the statement is never made explicit; but I think the book makes it clear. Krieger's implicit conception of the task of the social scientist is to represent life through images that generate more complex and interesting meanings and explanations than could be imagined by the writer. It is commonplace to picture theory in social science as such an open metaphor, as an evocative approximation to the confusions of life. It is less conventional to see simple empirical description in the same terms. I say "simple" with caution. I mean the simplicity of poetic description, a concatenation of suggestion built on precise craftsmanship. It would be unfair to threaten this little book with a comparison to poetry, but it can be read in the spirit of a prose poem. Krieger enhances our knowledge not by giving us a theory of organizations but by drawing us into a more elaborate appreciation of the events she describes, an appreciation built on her ability to record the fine detail of organizational life.

The fine detail is a record of some lives that crossed briefly at a radio station. At the same time, it is an examination of the ways individuals, institutions, and societies coexist and decorate that coexistence with moral crises. Individuals adapt to organizations; organizations adapt to the individuals in them. Organizations adapt to society; societies adapt to the organizations within them. Parents adapt to children; children adapt to parents. And so on. Depending on the discipline and ideology of the writer, such mutual adaptation is called variously learning, socialization, maturation, cooptation, or other similar terms, and is viewed as a sign of social intelligence or social corruption. It is a way in which the young and the boisterous are made older and tamer, a way in which the older and wiser learn from the younger and fresher.

The literature on mutual adaptation is primarily a recitation of two facts: mutual adaptation is ubiquitous to human existence, and it appears to be morally significant. Some people find it attractive, emphasizing the harmony and change it produces; other find it appalling, emphasizing the ways in which individuals are induced to conform to unfortunate social pressures: still others find it appalling for the opposite reason, emphasizing the ways in which traditional social structures are undermined by accommodation to deviants. The literature is distinguished by numerous studies demonstrating that adaptation occurs and numerous efforts to discuss the moral issues. Attempts to understand the microprocess by which adaptation takes place or the social orchestration of cooptation are less common. We know that it happens; we know that we do or don't like it; but we do not have an intimate sense of how it occurs. This book is a description of that microprocess in one case. Simultaneously, and mostly reluctantly, a group of radical rock radio performers are transformed into almost respectable professionals, and a pair of radio stations, a large corporation, and a governmental regulatory agency are made into slightly more progressive institutions.

The study is also an essay on social movements. During the late 1960s and early 1970s a collection of deviant themes — drugs, rock music, organic food, new left politics, sexual freedom, long hair, civil rights, alienation from the work ethic, egocentric psychology, and blue jeans — combined to form a loose coalition of people and ideas that came to be a powerful social movement. The movement was associated with changes in politics, music, clothing, marriage, beliefs, habits, and social philosophy. Birth rates, economic conditions, and the international balance of power were also significant to the changes; but there was a movement, and it affected society before being absorbed into it. Krieger's research examines how the movement and the society came to terms in a microscopic corner of the world. In the world in which real people live, of course, there was no "movement" and no "society." Those are fictions more useful to visitors and observers than to participants in life. Nevertheless, both the movement and the society were symbolized. There was a group of performers representing and supported by (not always consistently) the hip people and their media groupies; there was a group of radio owners and executives representing and supported by (not always consistently) the straight people and their bureaucratic functionaries. They met in one of innumerable small encounters that became, retrospectively, the history of a decade. Krieger gives a rich picture of how those encounters reflected and created the spirit of the 1960s and the process by which that spirit was coopted.

The usual explanation of mutual adaptation emphasizes rational exchange between individuals and social institutions. In crude form, it is assumed that rich and powerful institutions use their resources to bribe poorer and weaker individuals into acquiescence, that the exchange occurs because it is mutually beneficial in terms of the immediate self-interest of the parties. Because the parties rarely start from equal positions in terms of resources, the exchange rarely leaves them in equal positions; and the result is no more attractive than the intitial distribution of resources permits. But the presumption is that all parties act in a self-consciously rational way in pursuit of self-interest, doing the best they can.

The history reported by Professor Krieger has elements that are concistent with such a broad perspective. Bribes were offered and accepted.

But it is hard to find anyone in the story who could plausibly be described as consistently acting to maximize personal or organizational self-interest. They tried to make marginal improvements in their condition; they cared about their own interests; but they tended most of the time to take most of the situation as given and to drift into strategic actions rather than move into them on the basis of detailed planning or systematic analysis. They considered only a small number of alternatives and only a small amount of information about them. They used trial-and-error method. Some of them could be ruthless and devious, but their ruthlessness and deviousness were not particularly related to rational calculation. They were more frequently tied to impulse, pique, or habit. The owners, both individual and corporate, were concerned about profits; but they persisted in investing their money in a business with unclear prospects for a competitive return. The workers were concerned about jobs and wages; but they persisted in working for a marginal firm that paid poorly, and sometimes not at all; and on the way to the bank, they persistently stumbled over their dreams.

Without challenging the fundamental thesis that, in the long run, cooptation reflects an exchange of favors and allegiances in the interest of self-aggrandizement, Krieger shows some of the many frictions in that process that make its short-run course much less clear. Without denying the force of grand social explanations, she dramatizes the way an organization is a cross-section of individual lives, and every historical event an intersection of biographies. By exhibiting the complex context of organizational history, she reminds us of the difficulty of interpreting organizational outcomes as the necessary consequences of strictly organizational events. However, her main contribution is not in explaining why cooptation occurs or what the final equilibrium will be. What Krieger does particularly well is to show how this powerful social process is orchestrated, and particularly the role of outrageous behavior in the engagement between individauls and an organized society.

At a personal, organizational, and social level, this history is a morality play performed as a farce. It is a morality play because the participants saw it and acted it that way. It is a farce because the most obvious social rules were treated as arguable within a general context of social orderliness. The people who performed on the station did outrageous things. They were familiar things, though rarely said or done on the air; and they were things that were important to the rock music subculture. But they were unambiguously outrageous. They disturbed public sensibilities, and would have been otherwise almost pointless. Broadcasters were obscene in language and behavior, but their greatest obscenity was not in their actions but in the moral posture they adopted with respect to those actions. Socially acceptable sinners ask forgiveness for their outrages, claiming human weakness. These sinners asked for praise, proclaiming their outrages as a basis for pride. They challenged a rational society at a vulnerable point, its arbitrary beliefs in traditional decency, by attempting to convert articles of faith into subjects for rational discourse and by attempting to force moral and behavioral consistency on individuals and society. By asking what makes an obscenity obscene, they threatened the moral order.

It was not an idle threat. Society is built on those elements of judicious hypocrisy that permit the simultaneous encouragement of virtue and tolerance of sin, and on confidence that some things are proper simply because they are proper. Hypocrisy and faith are always vulnerable, caught as they are among the temptations to purity felt by virtuous people, the temptations to license felt by sinners, and the temptations to consistency felt by intellects. When it is possible for outrageous people to assert, without inducing laughter, that it is not they who are outrageous but conventional social rules, the farce becomes serious and we elaborate a new version of virtue. Stealing becomes restitution; earning is turned into theft. Murder is seen as execution, execution as genocide. Politeness is insult; crudity is honesty. Success converts to failure, failure to martyrdom. They were exciting years, and they left their mark on us all; but a view of virtue that requires moral rules to be completely consistent and completely defensible undermines the idea of a moral order in a way that cannot be sustained. And, of course, it wasn't. A new faith grew around a new rhetoric that allowed a retrospective enthusiasm for the outrages of the past and a commitment to new definitions of human dignity. Not the same as before, but not different either.

Krieger details the story of that struggle over morals in terms of the lives and feelings of ordinary people and ordinary institutions trying to get through, one step at a time. There are bureaucrats sitting in Washington wondering what to do with complaints about plain language on the air and thinking that maybe a trip to San Francisco is needed, particularly in January. There are managers sitting in New York trying to guess what the new wave in music is going to be and hoping they will have some hip insights when they go home to do battle with their children. There are owners of boutique shops with a counterculture clientele trying to pay the rent and find a cost-effective advertising program that doesn't make them into fat cats. There are disc jockeys, engineers, advertising managers, record companies, rock groups, newspaper writers, and each with family, friends, lovers, favorite metaphors, and time payments. And there are sentiments, sometimes decorating mundane activities with heroic significance, sometimes confusing personal mood with social condition, sometimes burying cynicism in romance, sometimes enjoying the trite comforts of pop philosophy, and sometimes achieving the small intelligences of folk aphorisms.

Not many organizations have a cast of characters comparable to the cast studied by Professor Krieger, or the special complications and exhilarations produced by the culture, ideology, and economics of rock music. Not many periods in recent history can claim the significance of the late 1960s for shaping the attitudes of contemporary generations. It is possible that this study has little relevance beyond that unusual period and should be read as an historical footnote, or as a novel. It is both of those, and the start of a film script as well. But though the book can be enjoyed in any of these ways, and should be, such enjoyments should not obscure its importance as research. It is a nice piece of sociology, with the skill of observation, awareness of detail, and sympathetic understanding of the human condition that distinguish fine examples of empirical social science.

James G. March
Stanford University
Stanford, California

Acknowledgments

I am indebted to all those persons whose names appear in this account. They gave of themselves as they spoke with me and trusted more than I deserved. I am especially indebted to Tom Donahue who died of a heart attack in 1975, eight years after he started the station. No one could defend the radio station as convincingly as Tom or reconcile its contradictions with as much style. He encouraged my doing a history of the station with a seriousness and inclusiveness that helped throughout.

I am grateful also to Martin H. Krieger and James G. March, each of whom gave early and continuing support, seeing value in a study that was at first by far too long. They met my stubborn insistence on detail and doing it all my own way with a belief in me independent of it that was at least as strong. Rita J. Simon at a critical point urged that I cut the orginal manuscript down to size and assumed that I was able. In addition there were people at every turn who were interested in the study, the story, the process, the station, and over a period of six years this was vital.

INTRODUCTION

In the spring of 1967, a former top-40 disc jockey met with the small businessman-owner of a radio station in San Francisco, California and arranged to take over a portion of its air time and devote this to rock music programming, expanding the rock music programming with added staff as soon as that became feasible. The station had previously had a foreign language format.

Several months later there was an influx of hippies and youth into San Francisco and the station, which had by then gone nearly full time with the rock music format, became known as an underground station, a bit of fancy, "flower power on the air." It was, according to one reporter, the only station in the world where:

> you'll see hips, frozen in the lotus position in the lobby. The Haight comes there every evening to chat, to ask for announcements to be put on the air, or listen to the music and use the crayons and paper supplied by the management for itinerant speed freaks who have nothing to do with their hands.[1]

According to another:

> Always its tone was the one in which we most deeply believed business should be done: taking it easy. KMPX's trademark: the programmer or someone fucking up, blowing something matter-of-factly up front on the air, backed by a chorus of giggles from the bird engineers. I remember, after one song an announcement: "It's okay, Rusty, wherever you are, it's not coming in on the plane tonight, you can relax. This is a public service." Silence. Then a shocked voice from the background: "But you can't say that on the air!" "Well, guess I just did," and Procol Harum flicked on.[2]

By the fall the station was doing increasingly well, gaining listeners and sponsors and showing up in the ratings. Three western-style salesmen who had been dope dealers had begun to sell its time and were introducing themselves in advertising agencies as well as in the shops of hip merchants. In the winter one of these salesmen took a $1,000 loan from a friend who was a dope dealer and had to dig up the money from where it was buried in his yard in order to pay the staff. That week they were paid with "muddy money" and their checks did not bounce as was usual at the time. The problem, the sales manager said, was that "the money was coming in irregularly. Payday, however, was regular." The problem according to the top-selling salesman was that the owner and sales manager could not deal with success. They were grabbing pennies. According to one of the "chick" engineers, the problem was that the man who introduced the format (the "Big Daddy" of the radio station, Tom Donahue) was spending too much time in Los Angeles trying to get a sister station off the ground. A lawyer for the station's owner felt it was simply that these people had found gold nuggets in the street and all they could do was pick them up and throw them at each other.

The following spring, one year after the rock music programming had been introduced, after months of mounting tension between the owner and the staff, Donahue walked out on a meeting with the owner in which he was offered the program director position at one or the other of the two stations, Los Angeles or San Francisco, which one he could not recall. He had thought they were both his. Three days later the staff met and voted to go on strike. "With the same conditions that he had to quit," one of them said, "so have we all." Their strike lasted eight weeks. A picket line was maintained in front of the station all during that time. It was joined by various psychedelic volunteers and members of the hip and music communities. Advertisers withdrew their accounts from the station, musicians requested that their records not be aired, local bands appeared in concerts to benefit the striking group. Both underground and establishment newspapers covered the story in detail. When the staff went off the air and onto the street at 3 a.m. initially, a crowd of about 500 people had gathered and were dancing outside in front of the converted warehouse building in which the station was located. There were signs, pickets, a light show, the Dead were playing. The strikers formed a union of their own for purposes of the strike: the AAFIFMWW, the Amalgamated American Federation of International FM Workers of the World, Ltd., North Beach Local No. 1. The FM could stand for frequency modulation or free men.

At the end of the eight weeks, after repeated attempts at negotiation with the station's owner had failed, eight members of the striking group went back to work at another station promising to bring more of the others on. This new station (KSAN), located in downtown San Francisco, had previously had a classical music format and was owned by a large media conglomerate. It was tucking their tails they went, the top-selling salesman said: "We are selling out going over there and we knew it. The justification for it was it was better than the streets."

The staff continued under the new corporate ownership for the next

four years, increasing the size of their audience, drawing wider advertising support, becoming evere more professional and successful, all the while struggling to reconcile their antiestablishment roots with pressures for conformity with an establishment of which they were becoming part.
   In the spring of 1972, Donahue came back to be general manager at KSAN. He had previously been program director and disc jockey. After one month in the position he was fired by the president of the Radio Division of the corporation, to be replaced by a sales manager from another station owned by the corporation in Oakland. The staff staged an angry meeting of protest in which they defended Donahue's keeping the position and reflected on their past five years. The president of the Radio Division told them at the start of the meeting that the station had been a fantastic programming success but it had not been earning enough money to pay back the company's investment and this was the reason for the management change. One of the staff present, a talk show host who four years back had been one of the original western-style salesmen, responded with the following statement:

> You got a radio station they could sell for five times what they paid for it in this market today. The thing was a cheap shit piece of stuff that was picked up by a conglomerate, moved by Kluge [the corporation's president] up from a half price five years ago, and he's now got something that's the number one property in FM today.

A female disc jockey, who had been a chick engineer with the station at first, made clear general feelings for Donahue:

> We've worked with Tom from way back and we watched Tom grow. He started the whole fucking thing. This was his brain child, and nobody seems to understand that.
>   The thing that makes the people in this room go and want to boogie and go full tilt is a man named Tom Donahue. ... The cat brings us energy. We're all a bunch of fucking downers, man, and when Tom comes in we fucking get our shit together, because we know that cat understands the programming.

Another disc jockey who had been with the station since the start asked:

> A critical question: What do you think if this radio station, meaning the air staff, most of it, if not all of it, and the sales staff, most of it, if not all of it, walked out on the street, quit, and took themselves as a package to another radio station with the ARB book. ... What do you think would happen to this radio station?

The president of the Radio Division responded that this was not the earlier situation all over again: "This isn't KMPX again, we're going to let you work." The talk show host then summed it up:

> Everybody back East in this corporation considers that we are a bunch of lunatics who have taken over the fucking asylum. Now, right now, what we are seeing is a changing of the guard because they're not going to let the boss fucking loony run the bin.

A disc jockey who had been with the classical format finally suggested:

> You're going to end up firing us one by one, because some crony in New York is a big contributor to the Nixon administration. You might as well do it now man.

But the staff was not fired. The president of the Radio Division reconsidered his decision and reinstated Donahue as general manager. Before this, however, in the next day's programming, the disc jockeys on their shifts all played "Slippin' Into Darkness," recorded by War, which the program director said seemed to be the company theme song: "You've been slippin' into darkness, pretty soon you gonna pay." When word came in the afternoon that Donahue had been reinstated, different people around said, "We won." The disc jockey then on the air punched off his microphone and said he felt this was a more important victory than the earlier strike, "this was a real victory."

Not long after, the female disc jockey who had been a chick engineer at first said that although she felt Donahue knew what worked and she was glad he was back, in some ways he was still in the old days:

> He thought they would do the work for love, do commercials on their overtime and things like that. He did not realize they were in it now for the money. Street felt one thing she had learned was you had to be cynical in this business, you had to be cynical in order to survive. You were going to be fucked in the ass and you had to take the pleasure you could get from it. That was better than just getting fucked and being realistic about it. You had to turn the joke. If you went to lunch with one of the salesmen to get a client, you had to think it was on them. You had to believe you were a freewheeling spirit even if that was not really the way it was.

This is a story of a radio station which went from idealistic origins to commercial success in the years 1967-1972. It is also a story of a more general process common to many of our organizational and individual lives: the process of cooptation. In this process an organization once new and different and at odds with prevailing practice comes over time to adopt ways of a larger society which are viewed as corrupting.[3] The organization is said to have sold out, to have lost its original virtue. The following account traces the history of one radio station in order to make clear the imperatives of this process. It is divided into four periods: Beginnings, Legitimacy, Professionalism, and Renewal. In the beginning we find an organization with a limited base of support. Over time in order to survive it grows and takes on an increasingly broad range of commitments. In doing so it moves away from its original purposes. People who judge it by its past are disappointed and accuse its members of having sold out. There is, nonetheless, defense of the station and continuing debate as to whether it can in fact avoid a corrupting fate.

I began inquiring into the history of the station in the spring of 1972 at a time when it was doing well and Donahue had just come back to be its general manager. I conducted research for the next year, seeking to trace the process of the station's cooptation. The research consisted of interviewing people who had been involved with the station during 1967-1972, obtaining documentary evidence, visiting the station and listening to it.[4] Many of the people interviewed had worked for the station at one time or another over the five years. They were disc jockeys, salesmen, secretaries, managers, owners, promotion directors, engineers. The rest had less central relationships to it but, nonetheless, felt themselves part: advertisers, promotion men, listeners, corporate and union personnel, accountants, lawyers, journalists.

In the next two years I wrote and revised the detailed history of the station on which the following account is based. This was submitted in 1975 as a doctoral dissertation in Communication at Stanford University.[5] It ran to nearly 800 manuscript pages and was an attempt to describe and explain the cooptation of the station by piecing together stories from the interview and documentary reports. The style devised to do this was a multiple-person stream of consciousness narrative. My purpose was to have the mode of description of the narrative be seen also as a mode of explanation, a statement as to why the cooptation process occurred as it did.[6] The choice of style was influenced by one particular view of cooptation I had found early on in a study of a telephone company in which it was said of customer service representatives:

> Working in that job one does not see oneself as a victim of "Capitalism." One is simply part of a busy little world which has its own pleasures and satisfactions as well as its own frustrations but, most important, it is a world, with a shape and an integrity all its own. The pattern of cooptation, in other words, rests on details: hundreds of trivial, but human, details.[7]

I sought to trace the process in terms of its details, systematically structuring their presentation yet leaving room for interpretation other than my own. I sought to include facts of how people felt and thought and what they liked in a world where immediate circumstance seemed often more determining than demands of a larger society. The assumption made was that cooptation worked, in important part, on an intimate scale.

The original study was revised in 1977-1978 after two years of putting it by, the length shortened, the story made more dramatic, introductions to each chapter added, but the style and purposes of the original maintained. The account which follows represents that revision. At the start of each chapter is an introduction which identifies major changes the station undergoes in the time covered by the chapter. A summary of these changes by period appears in the conclusion. At the start of the account as a whole is a list of characters and a chronology of events which may be useful as a guide to the reader. Sources for all evidence used are to be found in the final sources section. It will be noted that real names of persons appear throughout. This is a history of specific people and particular events and their actual names matter. The cooptation process the account describes is not a simple one. It is the purpose here to use the radio station case to set it forth so that comparison with the experience of other organizations can be made. The radio station of the present case could be any organization in its early years having to adjust and grow in order to survive.

CHRONOLOGY

The following is a summary of prominent events in the station's history.

Part I: BEGINNINGS
Chapter
1 Donahue begins programming on KMPX (April 1967).
1 San Francisco Hippie Summer of Love (June-August 1967).
2 KMPX goes full time in rock format (August 1967).
3 Expansion to KPPC (November 1967).
3 First ratings (November 1967).
3 Troubles on KMPX and KPPC (January-March 1968).
4 KMPX Strike (March-May 1968).

Part II: LEGITIMACY
Chapter
5 Rock staff moves to KSAN (May 1968).
5 Paulsen becomes general manager (August 1968).
5 Klein hired as sales manager (August 1968).
6 Sullivan resigns as Radio Division president (November 1968).
6 Bear fired after messing up commercials (February 1969).
6 Crosby sells KMPX to National Science Network (February 1969).
7 Donahue leaves after wedding with Hamilton (April 1969).
7 Duff becomes general manager (April 1969).
7 Ponek becomes program director (April 1969).
7 Harris becomes sales manager (June 1969).
7 People's Park controversy (May 1969).
7 Woodstock (August 1969).
7 Local competition affects sales (August 1969).

Part III: PROFESSIONALISM
Chapter
8 Anti-Ponek staff meetings peak (October 1969).
8 Prescott replaced by Congress of Wonders (October 1969).
8 Young fired after Nixon death threat (December 1969).
8 Altamont coverage (December 1969).
9 Nisker resigns after TDA rally and riot (February 1970).
9 Donahue returns on weekends (March 1970).
9 Duncan becomes head of Stereo Division (April 1970).
9 Boucher becomes program director (May 1970).
9 Bensky fired after Jeans West incident (May 1970).

Chapter
10 McQueen hired as news director (June 1970).
10 Pigg goes to ABC (June 1970).
10 Janis Joplin dies (October 1970).
10 KMPX Collective lock-in and firing (October 1970).

11 First waterbed account (November 1970).
11 Great Oil Spill (January 1971).
11 Street becomes music director (January 1971).
11 Simmons hired as record librarian (January 1971).
11 Second union negotiations begin (February 1971).
11 Duff becomes West Coast vice-president (April 1971).

12 Fillmore West closing (June 1971).
12 O'Hair hired as program director (August 1971).
12 Skinner hired as sales manager (August 1971).
12 Foster and Kleiser bombing (September 1971).
12 Revolutionary communiqué controversy (October 1971).

Part IV: RENEWAL
Chapter
13 Dr. Hip obscenity incident (November 1971).

14 Gay and women's shows discontinued (February 1972).
14 KMPX ceases rock format (April 1972).
14 KSAN-KMPX retrospective (April 1972).
14 Duff resigns (April 1972).

15 Donahue becomes general manager (May 1972).
15 Staff meeting over Donahue firing (June 1972).
15 Final engineering meeting (July 1972).

CHARACTERS

The following persons play major roles in the radio station story. A complete cast of characters appears in the Appendix.

Bear, Edward (disc jockey, KMPX, KSAN).
Boucher, Paul (engineer and production director, KMPX, KSAN).
Campbell, Donna (business manager, KSAN).
Crosby, Leon (owner, KMPX).
Donahue, Tom (program director, KMPX; general manager, KSAN).
Dougherty, Tom (chief legal counsel, Metromedia).
Duff, Willis (general manager, KSAN).
Duncan, George (president, Metromedia Radio Division).
Dunlop, Doug (salesman, KSAN).
Gleason, Ralph (columnist, San Francisco *Chronicle*).
Hamilton, Raechel (business manager, North Beach Productions).
Harris, Whitney (salesman, KMPX; sales manager KSAN).
Hunt, Ron (sales manager, KMPX).
Johnson, Katie (engineer, KMPX).
Laughlin, Chandler (salesman, KMPX; talk show host, KSAN).
McClay, Bob (disc jockey, KMPX, KSAN).
McQueen, Dave (news director, KSAN).
Melvin, Milan (salesman, KMPX, KSAN).
Miller, Larry (disc jockey, KMPX).
Nisker, Scoop (news director, KSAN).
O'Hair, Thom (program director, KSAN).
Paulsen, Varner (general manager, KSAN).
Pigg, Tony (disc jockey, KSAN).
Ponek, Stefan (disc jockey and program director, KSAN).
Prescott, Bob (disc jockey, KMPX, KSAN).
Street, Dusty (engineer, KMPX; disc jockey, KSAN).
Sullivan, Jack (president, Metromedia Radio Division).
Young, Roland (disc jockey, KSAN).

PART 1:
BEGINNINGS

In March 1967, Leon Crosby had owned radio station KMPX for five years. Tom Donahue was looking for work. Donahue, at 38, was a large and imposing man. His hair was dark and full, combed back. He had a close-cropped beard. There was an air of command about him, a seriousness, a presence which suggested the world should be his and would have been if someone else had not already dealt the chances. Donahue had come to San Francisco in 1961 to take a job as a disc jockey with the top-40 radio station KYA, where he soon became popular as Big Daddy Tom Donahue, "here to clean up your face and mess up your mind." He subsequently became involved with concert promotion, record production, and the psychedelic music scene which was gaining prominence in the City in the mid-sixties. During the fifties, Donahue had been a rock'n'roll disc jockey on a pioneering station in Philadelphia and had been implicated in the payola commotion of the late decade. He had also been an organizer in Democratic Party politics in a suburban Philadelphia county.

Leon Crosby was 43, a small man, balding, with an attitude of wariness about him, an uneasiness, as if he were surrounded by troubles which might at any moment get critically worse. Crosby had settled in the Bay Area in the mid-forties after getting out of the Air Force and had subsequently become involved with several local radio stations as an announcer, engineer, and part owner. He went by the name Country Lee and Hambone Lee Crosby in his early days in radio. Dixieland music was his specialty. In 1962, he bought a principal interest in KMPX from the owner of the San Francisco Warriors, a professional basketball team, paying $147,000 for it. He ran the station, starving with it, he said, and twice changing its format in the years 1962-1967.

The station had a jazz format in the 1950s and was known by the call letters KHIP. KHIP was relatively successful in its time but it was not doing well when Crosby took over. He immediately changed the call letters to KMPX and switched the programming to a middle-of-the-road format. He changed the programming again, to foreign language, in late 1966 when the Bay Area's previously established foreign language station dropped its format. As a foreign language station, KMPX ran on time let out to independent producers. They paid minimal fees to the Crosby-Pacific Broadcasting Company, of which Leon Crosby was general partner, found commercial support of their own, and provided the station with Portugese, Spanish, Italian, and Armenian hours. The format also included programs by student announcers who paid for their time as part of training in the Crosby School of Radio.

KMPX was broadcasting in stereo with a power of 80,000 watts, nondirectional, in March 1967. Its transmitter was located on top of Mt. Beacon in Marin County at a height of 1,250 feet. Listeners could tune it in at 106.9 on the FM band. The station was operated out of a modernized warehouse building at 50 Green Street in a section of San Francisco just off the Embarcadero, in view of the piers, and close to the base of Telegraph Hill. These were good quarters. For a period before, Crosby had operated the station out of a hotel room with not much more than a turntable and a microphone as equipment. The hotel quarters were such that listeners could occasionally hear when the toilet flushed and when money was deposited in a pay telephone nearby, or so it was said. For reasons having something to do with his hardship in running the station, and something to do with a tenacity one might not expect from a man of his appearance, Leon Crosby had a profound personal attachment to KMPX. The strength of that attachment was tested a year later, in the spring of 1968, when Donahue and a majority of the KMPX staff walked off their jobs and threatened to take Crosby's radio station away from him. What follows is a story of how that happened and of what became of the station in the next four years.

Chapter 1

KMPX
(March-June 1967)

In the initial months people come together to set up the new station, working out agreements in terms of relatively simple exchanges, making tentative, short-term commitments. They act in accord with institutional and personal contexts, assuming roles that are relevant and known: disc jockey, owner, investor, salesman, advertiser, listener. They gain space for the station gradually and in a series of trials define its style.

In March 1967, Donahue called Crosby at KMPX. The station's telephone was disconnected. Donahue subsequently met with Crosby and with the KMPX sales manager Ron Hunt who was closely responsible to Crosby. They got together on several occasions in March and early April, meeting at the radio station at 50 Green Street and in Donahue's apartment on Alta Street in North Beach to discuss the prospect of Donahue's handling program time on KMPX.

An agreement between them was reached which was verbal and temporary. The parties to it were the Crosby-Pacific Broadcasting Company, Crosby's firm, and North Beach Productions, Donahue's company. They agreed on terms by which North Beach would provide programming services for radio station KMPX. Donahue would go on the air from 8 p.m. to midnight, Monday through Friday, starting on Friday, April 7. He would play a mix of contemporary music with an emphasis on rock. His show would replace a Chinese language program which was behind in paying its bills. North Beach would sell commercial time for the 8 p.m.-midnight period and receive a third of the gross amount booked. Donahue would expand the kind of programming he introduced into other time periods as soon as that became feasible, as soon as people suited for the work could be found and contracts of the foreign language programs then on the air could be terminated.

Donahue claimed North Beach Productions was never paid according to the initial agreement negotiated with Crosby-Pacific. Several weeks later, toward the end of April, a second agreement was made. It, too, was verbal. According to it, Crosby-Pacific would pay North Beach $150 a week for programming services and $150 a week as a nonreturnable advance against commissions made off time sold on Donahue's 8 p.m.-midnight show. Commissions would be 35% of gross sales.

There were two other people of interest present at certain of the initial Crosby-Pacific North Beach meetings. One, Lew Avery, had been sales manager and part owner at KYA when Donahue worked there. In April 1967, Avery was retired from professional media sales work and in his early sixties. He had money he wanted to invest and Donahue's venture on KMPX appealed to him. Between April 21 and May 12, he made loans to KMPX totalling $45,000. He made his loans payable to Crosby as a personal favor and received simple promissory notes in return. Crosby promised to return the amounts with 7% annual interest at an unspecified date of maturity.

The second, Raechel Hamilton, had been Donahue's secretary at Autumn Records, a previous business, and had joined him in the incorporation of North Beach Productions. She lived with Donahue and worked closely with him. Hamilton was 20. She had left home at 17 after high school in southern California, come up to San Francisco to go to school, and worked at secretarial jobs and in several bars in the City before she became involved with Donahue.

Donahue's 8 p.m.-midnight show was not the first of its kind to go on the air on KMPX in 1967. Crosby had hired Larry Miller in February to do the all-night shift on the station from midnight to 6 a.m. for $45 a week. Miller's show was described as comic in personality and eclectic in musical style. He biased it toward folk and classical music with some jazz and some rock. When Donahue began doing the 8 p.m.-midnight shift in early April, an unknown number of listeners had already heard Miller's show and associated the station with it. Miller had received a surprising number of favorable letters from listeners not long after he started. It was said Crosby did not believe the response at first. He thought the letters must be coming from Miller's friends and relatives. Then he realized Miller could not have so many friends and relatives. Miller was 26. He had come to San Francisco in 1966 from Detroit where he had been a folk music disc jockey for two years before.

Miller recalled there were about 50 records in the KMPX record library when he started in February. He brought in records of his own to play and went out to record stores and distributors asking for more. But he met with only limited success in expanding the station's collection. He was a nobody on an unknown radio station playing esoteric music in dead hours of the night. At one point he drew up a poster to publicize his show, had a 1,000 copies made and handed out many of them on Haight Street. He earned money in addition to his salary by selling commercial time on his show. Miller said he heard in March that Donahue was negotiating an agreement with Crosby but he did not know at the time what Donahue had in mind. He thought Donahue may have heard his show on the station and anticipated it could be more widely marketed. Donahue said he had not heard Miller's show before he talked with Crosby, he did not have an FM radio at the time, but he had been told about it. Ralph Gleason, a San Francisco journalist and music critic, remembered telling him.

On March 19, Gleason and Donahue took part in a panel discussion on rock'n'roll at a University of California Extension conference at Mills College in Oakland. At one point during the discussion, Donahue commented on the situation of rock music radio in San Francisco as he then saw it:

> There are certain albums that outsell singles, particularly in San Francisco. But you don't know about it because the radio stations don't have the guts to play them. I know Donovan has had albums in this town that have outsold single records, that should have been in the Top 10. But in most of these radio stations, the program people don't have the faith in their own ear. They are afraid to go into an album and play the guts out of it. They are afraid somebody will tune out if they play the wrong one.

Toward the end he spoke in more visionary terms:

> I think there will be a change some place along the line. Some radio station owner will have the courage to come up with a new kind of radio. The Top-40 radio format dates back to about 1955. Some people in Kansas City started it. Nothing has really changed since then, they are all doing things they have no idea why they do, because that's the way it's always been done.
>
> Some place along the line somebody will break it, will come up with an imaginative, creative format. And as soon as that's successful, 2,000 stations throughout the country will start doing that and wear it out in no time. But it'll change because it can't stay this way. It's too boring.

Gleason said after the panel discussion, he suggested to Donahue that KMPX might be a possibility for the kind of radio he had in mind.

Donahue did not recall the incident of Gleason's suggestion. He did recall a period during which he was talking to almost anyone who might be interested in his ideas for a new kind of radio. He recalled an associate in the record business in Los Angeles advising him to look into stereo FM. He was imagining a radio programming style which would be markedly different and perhaps antithetical to the style of Top-40. He had in mind doing away with the jingles and the fast talk and jive which had by then become standard in Top-40, still playing rock music and especially new releases, but selecting them according to personal taste and presenting them in a manner which assumed a cordial rather than a pressured rapport with the listening audience and which conveyed the sense of a one-to-one relationship. He recalled initially expecting to program from a playlist, designating selections to be aired beforehand. A friend asked why he would have to do that and he thought in response he would never find enough people he could trust.

Donahue, in March, was not exactly sure what he would do but he felt he was right about it. He knew things had never come easily to him, he had always had to hustle for what he got. He had been out of steady work for over a year and out of radio for nearly three years, but he was up to date with the music. He had business connections and relationships with bands, record companies, and producers. He sometimes played music he liked, especially new music, for friends in his living room in the manner of a disc jockey. Being involved in the rock music business meant relating to people considerably younger than himself. Donahue liked that. In the spring of 1967, he could look back on a series of personally important experiences with acid which he felt had affected his thinking about radio. He was separated from his wife, with whom he had four children, and he had been looking for work for several months.

Donahue had been inquiring into other radio stations before he called Crosby at KMPX. At first he was looking only for an announcing job, not the programming of an entire station. One of the stations he approached was KYA where he had worked when he first came to San Francisco. KYA refused him. He understood the station owners feared they would get in trouble with the FCC if they took him on. Although he had been a key figures in KYA's achieving success in the early sixties, and although he was known for professional competence, Donahue was suspected of corruption.

Donahue, in his own eyes, was not corrupt. But he knew there was enough in his past to feed the iminations of those who would distrust him. He had done criminal intelligence work in the Army and at one time had ambitions in the direction of the CIA. He had never fully renounced his association with payola practices in Philadelphia and he seemed to take pride in viewing himself as one of the "lepers of the industry." He was a grown man who had been playing music for kids before that had a cloak of intellectual respectability. He had never been content to commit himself exclusively to one job. In Philadelphia, he had worked a 6 a.m.-10 a.m. radio shift, then turned up at 11 a.m. at a Bucks County office where he worked as a township official under another name, his given name, Tom Coman. In San Francisco, he had moved from radio into related businesses with surprising agility.

Donahue, on several occasions, had taken steps to clear himself of allegations of wrongdoing. In February 1965, the FCC began an investigation of himself and a fellow disc jockey at KYA, accusing them of conflicts of interest because of their outside business activities. Two investigators from the Commission came out to San Francisco and made a six-week search for evidence of payola or plugola on their parts. They found none. Donahue followed up their visit with a trip to Washington in December 1966, met with the FCC chairman, and obtained a letter stating that the Commission's investigation had found no evidence which would incriminate either himself or his associate. That was official clearance, yet it failed to dispel the beliefs of those who wanted to distrust him. They seemed to know almost for certain that somewhere, near to the core, he was a dangerous man.

On Friday, April 7 at 8 p.m., Donahue went on the air on KMPX for the first time and received an encouraging number of phone calls. After his first week on the air, he asked listeners to donate things to decorate the studio: beads, feathers, and bells. They might bring them down or send them in. By the end of his second week, mail had begun arriving and people were stopping in at 50 Green Street with decorations. They brought beads, dope, feathers, posters. Donahue remembered them coming at night while he was doing his show. He let them hang around, he liked having company. He took the clock out of the studio as if to spite top-40, he wanted never again to announce the time. During his second week, he got calls from two friends at KYA who wanted to know what he was doing. They told him listeners had been phoning them asking to hear records he was playing on KMPX.

Donahue recalled initially expecting he would have to be on the air for two months before he could seriously sell commercial time. But then he got his first sponsor during the second week, The North Face, a ski and hiking goods store in North Beach. The manager of The North Face, a young man, called to say he had been listening to the station and wanted to buy time. Donahue said he told him he had not yet devised a regular schedule of advertising rates but he would sell him spots at $10 each, one a night on his program. On Friday morning, at the end of the week, the manager of The North Face called back to report that 21 people who came into his store said they had heard about it on KMPX. He then drew up a fact sheet for Donahue to use in a second week of commercials. The fact sheet said the manager of The North Face and his wife had started the business themselves in 1964. They were enthusiastic and semihip. The idea of their store was low-key, soft sell, with emphasis on letting people look, not bugging them. The North Face had expert sales help, top quality things, no cheapo merchandise, no dime store attitudes. Their second week of radio spots would offer a $1 discount on sailor shirts with the mention of KMPX listening to determine how effective the advertising was.

At some point during his first week, Donahue called Avery and asked him to listen. He had visions of Avery in his Stanford Court apartment with a radio in every room listening to KMPX, and Avery a Rockefeller Republican. Avery listened and consulted several associates, one of whom was media supervisor at the Lennen and Newell advertising agency in the City. She knew Donahue from KYA and told Avery she expected KMPX would reach a market people in the advertising business were looking for, the 20-34 age group, which was not being reached as a distinct group by any of the radio stations then on the air. She could see how KMPX might deliver that group, the ones who had outgrown teenage rock'n'roll and were not yet ready for background music.

By the end of his second week, Donahue knew he did not like the man who was engineering his program. KMPX was set up for two-man program operation. An announcer handled the music while an engineer handled the technical controls in a space divided from the announcer's booth by a panel of glass. After his second week, Donahue said he told Crosby he wanted to hire himself a new engineer. He announced on the air he was looking for one, and that he wanted a woman, a chick engineer. On top-40 radio the engineers were old men. Bob McClay remembered suggesting to Donahue that they hire female engineers on KMPX. McClay had taken a shift on the station at Donahue's invitation shortly after Donahue began. He was recently separated from his wife and in the process of falling in love with a waitress at the Minimum Daily Requirement in North Beach who he thought he might like to have engineer his show. He asked her. She was interested. She learned enough to take the FCC examination and qualify for a third-class license and then joined KMPX in May as the station's first chick engineer.

McClay, like Donahue, had a background in top-40 radio and, like Donahue, considered himself a professional radio man. He was 27. He had come to San Francisco from New Jersey by way of New York in 1965 to take a job at KYA. He had known Donahue at KYA and had worked with him in Autumn Records and Tempo, businesses owned jointly by Donahue and a fellow disc jockey at KYA, Bob Mitchell. McClay had agreed to purchase the Tempo newsletter and record service business from them and was spending his days running Tempo at the time he started at KMPX. He felt he had to do both jobs because he could not take KMPX seriously in terms of money.

McClay recalled Donahue had spoken to him about KMPX when he was first meeting with Crosby in March. Donahue knew he was tired of the formulas of top-40 radio. Donahue then called him in April when the 1 p.m.-4 p.m. shift on the station opened up. It was followed by a Japanese program and then by Donahue's show from 8 p.m.-midnight. McClay remembered Donahue telling him he could play and say whatever he wanted if he came to work at KMPX. In McClay's experience, that was extraordinary opportunity. He had worked for a while at the top-40 station KFRC in San Francisco after he left KYA and among the strict format rules introduced at the time was one requiring the disc jockeys not to announce their names.

When Donahue announced in April that he was looking for a chick engineer for his program, Warren Van Orden heard him. Van Orden listened to the radio in a way few people do. He noticed slight changes. He memorized formats, especially those of rock music stations. He would notice the introduction of a new line in a commercial or a jingle, an irregularity in the scheduling of a program or the style of a disc jockey, a new sound in place of an older one which, too frequently repeated, had become boring and been abandoned. Van Orden worked as a volunteer at KPFA, a noncommercial radio station in Berkeley. He told another volunteer there, Katie Johnson, that Donahue was looking for a female engineer on KMPX, she should call him.

Johnson had first learned engineering on a radio station at Pomona College in southern California, had subsequently transferred to the University of California at Berkeley, and had then started working part-time at KPFA. In April, she ws producing children's programs and poetry and jazz collage programs for KPFA, working with a friend, the two of them excited by what they could do with radio sound equipment. Johnson remembered calling Donahue early in May. He asked her to come over to see him at the station at 8:30 the following night. She turned on KMPX afterward to hear how it sounded. A Portugese program was on, followed by music she had not heard before. When she went to see Donahue the next night, the building at 50 Green Street impressed her as fresh and new compared with the offices of KPFA. The walls of the radio station on the second floor were covered with feathers and other decorations. She was told that listeners had sent them in. Donahue said he had been interviewing dozens of candidates and she was the first to come with previous radio experience, then he offered her the job as his engineer. She agreed to start on Monday after school ended for her at Berkeley.

It was probably on Monday, June 5 that Johnson went to work as the second chick engineer on KMPX. She was scheduled to do the 6 p.m.-midnight engineering shift for a $1.50 an hour six days a week. For the first two hours each weekday, she would be engineering McClay's show, moved from its earlier time. Then she would do four hours of Donahue's show, and on Saturday nights, Voco's show. Voco, a friend and neighbor of Donanue's and an enthusiast of the blues, was regularly employed as a branch manager for Mercury Records. Johnson said she found the 6 p.m.-midnight engineering shift tiring but she was 21 then and devoted to radio. She believed she was practically the only person at the station who had not come through friends. She thought of herself as having come in out of the blue. She had experience and perfect pitch, but she was not familiar with the world of professional music or comfortable with drugs. She learned about both from Donahue. Donahue programmed his show in sets of three. He would play three related cuts and between sets discuss the musicians and what they were doing.

Through the late spring and into the summer, Johnson played records for Donahue, McClay, and Voco. The records were from the KMPX library which was stocked from the personal collections of the three of them. It had several hundred records by late June, catalogued by subject and color-coded by owner. Johsnon also played the commercials on each show. The station's equipment for commercials consisted of two home tape recorders. The commercials had been recorded on tapes of different lengths and levels, with different types of leads, and the tapes themselves were of unmatched kinds. Johnson thought the equipment they had at KMPX was worse than at KPFA. The equipment at KPFA was at least set up according to radio conventions. At KMPX, what should have gone left to right went right to left on the control board, parts were missing. Johnson assumed the inadequacy was an inheritance from Crosby's radio school and felt obliged to try and fix it.

She also answered the phones. Eight telpehone lines came into the station. It seemed to Johnson they rang all the time. It was the Summer of Love in San Francisco in 1967 and many of the calls were from kids spaced out on drugs. They wanted someone at the station to hear them and talk them down from bad trips. Johnson remembered she could not handle the drug calls, but Donahue could and sometimes did. There were a lot of calls for help. Once there was someone who called and said he and his girl friend had been on an acid trip on Montgomery Street and he had lost her, what should he do? There were calls from people complaining about the station's commercials, calls to say, "Wow, I like your station," calls criticizing the way Donahue ran his records together. There were requests for announcements about lost dogs. Doctors and dentists phoned to say they used the music in their offices. A few people in peculiar circumstances called repeatedly. One, a girl in a hospital with gangrene, called on and off for three months. Johnson said finally she turned her over to Tony Bigg, a disc jockey who did the 4 p.m.-8 p.m. shift on KYA and often came over to KMPX afterward to hang around. Bigg talked with the girl in the hospital one night. A while later she showed up at KYA in her bathrobe to see him. He brought her to KMPX and the hangers-on at the station took her in. Johnson remembered that episode and she remembered how, starting in the beginning and continuing through the summer, people came and shuffled through the radio station at 50 Green Street at night, most of them spaced on drugs.

Bob Prescott said he first heard of KMPX during the second week in May when some girls he met in Golden Gate Park told him about it. He listened to Donahue's show one night and thought he was hearing the kind of radio he had been talking to fellow disc jockeys about for years, but which he was convinced would never happen unless one of them inherited a radio station out of the blue. Prescott was 30. He considered himself a professional radio man and a regular worker in radio, not a star. He had been out of radio work since 1965. A succession of experiences had left him disillusioned with the radio business, the politics of it. Prescott had dropped out of high school in Hawaii, joined the Navy in the mid-fifties, then worked at a variety of radio jobs, at top-40 stations, soft rock, middle-of-the-road, and automated stations. He had come to San Francisco in 1967 after spending a year in New York City as manager of the Cafe Figaro in Greenwich Village.

Sometime during the third week in May, after he heard Donahue's show, Prescott went over to KMPX at 50 Green Street, made an audition tape, met with Donahue, and tried out on the air on the weekend. Donahue arranged to hire him full-time beginning in the end of the month.

Milan Melvin in the end of May was dealing dope in San Francisco. He was 24 and sharing an apartment with a friend who was an actor. He had never before in his life worked for a radio station. His friend knew Donahue from a time when Donahue had run a psychedelic nightclub in North Beach called Mother's. Melvin thought his actor friend may have been the one who first directed his attention to KMPX and recommended he go see Donahue. Melvin recalled listening to KMPX and liking it, then going and telling Donahue he would do anything, sweep the floors, anything, to get near that radio station. Donahue told him the station needed money. Donahue said he thought if Melvin could sell dope, he should be able to sell time.

Melvin remembered going to work for KMPX in May with a revolutionary zeal. He knew that radio time salesmen ordinarily worked for commissions, earning a percent of the amount of sales they billed. In this case commissions would be 20%. But he did not think of himself as working for commissions or for money at all for KMPX. He could make more dealing dope. The incentive he remembered was dream power. He recalled hitting the streets hard beginning in May, going to every hip business he could find, telling them KMPX was a possibility which never existed before. The businesses he approached at first were boutiques and head shops which did not do much advertising, and if they did, were not likely to have advertised on radio before or to have given it thought. They probably had not even heard of KMPX. Melvin said his first task was to sell them on listening to the station. He would bring in FM sets for them to use. He had to promote FM radio and the merits of the KMPX approach before he could get down to selling time.

Melvin felt his purpose as a salesman at first was just to get advertisers on the air. He had no evidence to prove it, but he was convinced once they were on they would get results. He went around dressed like a bohemian with an earring in one ear, long dark hair, sometimes a purple cape, a top hat or beads. But he talked business and impressed those he spoke with as genuine. He made deals, trade-outs, half trades, he gave time away. One of the shops he went to in the beginning was Mnasideka's, a clothing store on Haight Street. He thought he had had almost sold the manager of Mnasideka's on the station. He had gotten her to turn her radio to it. She was listening. He left her listening, came back over to 50 Green Street, and asked Donahue to plug Mnasideka's on the air. Donahue plugged it, the manager of Mnasideka's bought time, and Melvin was out on the street again. One day he stopperd into Music City on Columbus Avenue in North Beach and told the owner and manager if he would give him the violin case lying on the counter in his store, he would sell him time on KMPX at a cut-rate. The owner and manager of Music City recalled feeling at the time it was a little strange. Melvin said he wanted to use the violin case as a brief case. He told him he would think it over.

In the absence of ratings which would show how the station was doing in terms of numbers of listeners, Melvin carried around letters from listeners in his pocket to show to potential advertisers. One, written April 26, read:

Dear Underground,
Your radio station is altering our entire domestic routine. For one thing, I customarily listen to the radio when I perform my morning ablutions, prior to going to work. The radio informs me of the time, whether or not it is raining, whether or not World War Three has begun, and other bits of relevant information. Now that I am able to listen to Underground Radio every morning, I find myself completely losing track of time, oblivious to changes in climate (physical and international), and so turned on I am unable to turn off in order to make the change from personal and private to public and businesslike. The Ecstatic Experience flashing through my ears has led me to flamboyant changes in apparel and make-up, so that now I arrive at work in outfits ordinarily reserved for weekend day-tripping.
And later than usual, too.
And then in the evenings, well, my god. My old man and me used to enjoy dining out, going to the Matrix, going to visit friends, taking long walks, or simply spending a quiet evening at home watching TV and making out. Dinner out is impossible, of course, because of the necessity of being home in time for Big Daddy. TV is out, naturally. Nobody comes to see us anymore because of the vows of silence and attentiveness required at the door. Making out is permissable only if done quietly and without too much distraction.
And then, because you stay on the air all godamn night, and as a working couple we've got to sleep sometime, we have had to purchase an extension speaker for our rather immobile stereo/FM unit to put in the bedroom. And getting lulled into sleep by The Velvet Underground doing "Heroin" and The Doors doing "The End" gives rise to subconscious psychic disturbances which manifest themselves into strange, awesome and occasionally terrifying dreams.
It's a mindbender ok. And if your underground programming were to become full-time, without the sane intermittent lapses into Carlos Albuquerque, I imagine my career, our domestic tranquility and my health would be completely shattered.
A captive audience of two,
SL and BC

Chapter 2

KMPX
(April-November 1967)

In these first eight months, different kinds of interest in the station feed each other. As the station and its audience become known, advertisers get results, rates can be raised, the station income increased, new people taken on, performance improved, record companies attracted, more listeners gained. There is confirmation for early investment and encouragement to do more. The very facts of success are dramatic enough for public story.

In mid-April, Crosby began inquiring about purchasing a second station in the Los Angeles area together with Avery. On April 14, a media broker wrote to Crosby asking him to plan a trip to Pasadena to look at radio station KPPC, AM and FM, owned by the Pasadena Presbyterian Church. KPPC operated out of the basement of the Church in downtown Pasadena on East Colorado Boulevard, with its transmitting antenna located next door on the roof of the Pasadena newspaper, the *Star-News*.

Crosby's lawyer recalled one afternoon in mid-July, Avery came with his son or son-in-law, a lawyer, to meet with Crosby in the office of Crosby's lawyer in downtown San Francisco. Avery agreed at that meeting to put up all the money required for the purchase of KPPC, to loan Crosby capital for a 60% share in the ownership and to himself invest directly in a 40% share. On July 13, Crosby signed an initial purchase agreement with the Pasadena Presbyterian Church for the sale of KPPC. The FCC granted authorization for the sale on October 5 and Crosby and Avery joined in a corporation, the Crosby-Avery Broadcasting Company, which assumed control of operations on November 6.

Donahue was impressed by the speed with which Avery and Crosby managed to get FCC approval for their purchase of KPPC. They received authorization for the sale 60 days after filing for it. It seemed to Donahue that Avery's contacts paid off. Donahue looked forward to expanding the format to KPPC. He believed he and the others from KMPX would be spreading a good thing by programming KPPC on the model of KMPX, when they might have kept it all to themselves in San Francisco. But to some of the staff of KMPX, the expansion seemed also a

43

test. They thought of Los Angeles as a tougher market than San Francisco and they hoped with KPPC to show that the success of KMPX was not a phenomenon limited to San Francisco or to the summer of 1967.

The success of KMPX, such as it was, was owed in part to an engineer named Paul Boucher. Boucher had come to work at the station in early July as chief engineer and production director. The first title meant he was responsible for keeping the station's equipment up to FCC regulation. The second meant he produced the commercials. Boucher recalled buying a modulation indicator for KMPX soon after he arrived, part of an effort to bring the equipment up to code. He spent $2,500 on a tape machine to be used for producing commercials, but mostly he negotiated trade-outs: four loudspeakers from Skinner, Hirsch and Kaye, cartridges and needles from the Stanton Company, earphones from Koss. In arranging the trades, he avoided making a big deal of the change taking place at the station.

As production director, Boucher recalled in the first four months taking commercial copy sent in by advertisers and changing it. When it was top-40 copy, he tried to tone it down and make it clever. In some cases where a spot said "we," he changed it to "they," feeling this would help keep the station's authority separate from that of its advertisers. By September, he had begun producing spots of his own for standard products like Leslie Salt and Mother's Cookies. These were for the station's salesmen to present to advertising agencies handling the accounts as examples of what KMPX could do. The spots were turned down by the agencies, who Boucher said were committed to seeing that a client's commercial sound came across the same on all air. But for Boucher there was still satisfaction in changing or sending back the copy. As people in the agencies saw it, he thought, a radio station refusing to run their copy was like a printing press refusing to print.

In July, several local theater people had a hand in producing some of the KMPX commercials. A few actors from the Committee came in to do spots in exchange for time on the air to publicize the New Committee Theater. The Congress of Wonders, a three-man comedy team, did a series for Far Fetched Foods in return for a supply of health food which Boucher thought must have lasted them a year.

Boucher said he initially objected to the hiring of chick engineers on the station. It seemed to him an unwarranted expense. But later he changed his mind. Boucher was 39. He had started in radio as an operator in the Merchant Marines. In that way, he thought, he was like David Sarnoff, the chief executive of RCA. He considered himself a radio man and by that he had in mind a curious, useless breed of person.

He had worked as a country and western disc jockey on a station in Needles, California after leaving the service, then at KFRC in San Francisco for three years where he got experience with live music engineering, then he moved to KYA where he worked as an engineer for 12 years.

Boucher knew Avery from KYA and from what he knew in the beginning at KMPX, Avery hired a number of salesmen and encouraged them to approach potential clients in a customary business manner like the salesmen he had managed at KYA. They failed to sell the station. Donahue did not speak of that. He liked to recall how Avery devised the first rate card they used at KMPX. It went into effect May 15, raised previous advertising rates and changed their structure. In Donahue's eyes, it was significant to have a rate card devised by Avery. Donahue believed that before Avery came to San Francisco, the radio stations all butchered their cards and undercut themselves by selling for whatever they could get. But Avery would not do business like that.

A schedule of commercials for Gramophone Shops first ran on the station in prime time early in June. A schedule for Music City first ran early in July. A schedule for the Bead Freak began the last week in July. Each of these advertisers felt they had a special relationship with KMPX. The owner and manager of the Bead Freak went on the air on July 24 to publicize the opening of his store at 511 Irving Street. The following week, he wrote to the station enclosing a check for his spots and thanking the staff, Miller and Melvin especially, saying he was indebted to them for his good start in business. The owner and manager of Gramophone Shops wrote on June 28 that advertising with KMPX had increased the demand for English LP's to such an extent that he had to reserve cargo space twice weekly on London to San Francisco flights. He asked the station's disc jockeys to thank their listeners for their response and to request their patience if his stores were temporarily out of stock.

The owner and manager of Music City, who had his doubts when Melvin first walked into his store and asked about arranging a trade-out for the violin case lying on his counter, had been a drummer before he got into the business of selling instruments to local bands in 1965. He considered himself still new in business and was cautious. He listened to KMPX after Melvin's visit and heard Eddie Kramer's Music World advertising on it. He called Kramer and asked him about KMPX. Kramer told him the station was in with the local musicians, it played tapes and long cuts and its spots were done on an ad lib basis. He then got back in touch with Melvin, gave him a fact sheet on Music City, and signed a contract for a schedule of commercials to begin the week of July 3. He later wrote a letter for the KMPX salesmen to use in promotion saying the monthly gross of Music City doubled in each of the first three months after he started advertising on the station and he had cause to move Music City to larger quarters, a renovated automobile showroom at 817 Columbus Avenue. During those months, he remembered, several of the KMPX disc jockeys came over to talk with him at his store. In his mind, his business and that of KMPX grew hand in hand.

As soon as the station began airing a variety of commercials, listeners began discriminating among them and making their preferences known. On June 23, one listener in Berkeley wrote that he thought the commercials Donahue read on the air were sometimes humorous and pleasant, but that the station had several produced commercials which were downright offensive, notably one about tranquilizers and one about vitamin pills that came packaged in a plastic owl. On July 3, Donahue replied that he agreed with the listener's feelings regarding some of the station's commercials, but that the station had commitments negotiated prior to the new format and was obliged to see them through. He expected certain potentially offensive commercials would be rewritten and produced in keeping with the new format. But in the meantime he hoped the listener would understand the station's dilemma and bear with them through a few more vitamins.

Donahue also wrote to record companies at the time, requesting copies of records and inviting visits to the station. He recalled the companies varied in their generosity. Elektra offered to give the station anything they wanted beginning in June. Vanguard also recognized them and supplied them with records in the first few months. But others were reluctant and Donahue had to take the initiative, making use of his contacts in the companies generally, looking up people he had known as promotion men who had since made it up executive ladders. The record library in the summer depended on McClay who got records through Tempo, and Voco who got new releases through Mercury. They each sought out people in distributorships who were willing to take a chance.

On Sunday, August 6, KMPX finally went full-time in the rock music format Donahue had introduced in April. The foreign language programs had all been edged out but for a few remaining on Sundays. Filling out the weekday schedule were people Donahue had recruited, some Melvin had recruited, and some who had just turned up. Melvin had found a third chick engineer, Dusty Street, who came in August to complete the regular engineering schedule. The regular announcing schedule was filled out with one more former top-40 disc jockey. There were part-time people on the weekend shifts. Johnson had brought her friend from KPFA. She and two other chick engineers worked weekends and did relief. Melvin's actor friend did a show on Sunday afternoons. Another actor from the Committee did a Saturday afternoon shift. Voco was on Saturday nights. The weekend people were paid at the same rate as the chick engineers, $1.50 an hour. Voco recalled not taking a cent. The fulltime announcers made $100 a week, except for Donahue who was paid for programming services. Donahue's eldest daughter worked as the station's receptionist at the part-time wage, his eldest son worked as janitor.

Melvin was the top-selling salesman through the summer months. There was one other regular salesman and Hunt. Melvin recalled he and Hunt tried a number of salesmen during the summer but they did not work out. Then toward the end of September, he recruited two who did, Chandler Laughlin and Jack Towle, both friends of his from several years back. They had known each other from dealing dope. Towle had been one of the founders of The Family Dog in the summer of 1965 and was said to have helped get The Family Dog starting capital by purchasing a lid of grass on a loan from Household Finance.

In September 1967, Towle was doing booking and promotion for The Western Front, like The Family Dog, a rock concert and dance agency. Towle said he was bailing out of The Western Front when Melvin came to him and asked if he wanted to sell radio. He told Melvin he had not sold anything legal before. Towle was 27. He had done some radio work in college and electronics in the Navy. The second salesman, Laughlin, was in jail in Contra Costa County on marijuana charges when Melvin first sought him out. Melvin had Donahue send a letter to help get him released and in the end of September, Laughlin was there at the station with a right-out-of-jail philosophy, wanting a new pair of western-style boots. Laughlin had been in business with Melvin before in a pottery and jewelry trade with some Indians in the Southwest. Their business paid for the Indians to manufacture pottery which they sold back to the Indians in exchange for silver jewelry. They then shipped the jewelry east to Provincetown, where it was bought by tourists.

Laughlin had known Towle from The Family Dog. Prior to the first concert and dance of The Family Dog in 1965, Laughlin had worked as a bartender at the Red Dog Saloon in Virgina City, Nevada. In Melvin's mind, the Red Dog Saloon was Laughlin's idea of how to reactivate the West. Laughlin had arranged for the owner of the Red Dog to hire The Charlatans, a newly organized western-style rock band, which according to Gleason was the first San Francisco band, and after a debut in Virginia City, the Charlatans appeared in San Francisco at The Family Dog. The lead singer with The Charlatans at the time was Laughlin's old lady in September 1967. A month after Laughlin joined KMPX as a

salesman, she came to work at the station as sales secretary.
Melvin said when he and Laughlin returned to San Francisco from Nevada in 1965, they considered going to radio school and visited one in the basement of a hotel where they met a man Melvin later thought may have been Crosby with a wig. Whatever the case, he and Laughlin were in earnest when they went down there, but when they saw what it was like they decided it was too greasy and left. Melvin continued to feel Laughlin was a radio man, although Laughlin did not have training or experience. He fantasized radio shows in his head, he would talk them out loud in the bathtub. Laughlin was 30 when he came to work at KMPX. Politically, he thought himself psychedlic right wing. He was against big government, against big business, an individualist, more conservative, he felt, than his father.

Towle, too, was conservative. He could easily become indignant on the subject of Communists or Cesar Chavez taking people's money. Towle, Melvin, and Laughlin said they kept their dope dealing contacts while they were selling time on KMPX. They did it to stay alive. Of the gross amount they billed for the station each week, not more than half would be collected and then not for a while. As salesmen, each of them made 20% on what he sold. That was not much compared to what they could make off dope. Melvin believed he was selling time on KMPX for love, because it certainly was not for money.

Laughlin recalled what he did to sell the station at first was to walk into a shop he thought catered to their type of people and tell the manager KMPX had the 18-34 audience, the ones who would buy his products. The manager would see in the next rating book that the station had the audience, although it had only been on six months. If he wanted to know sooner, he could ask the first five people who walked into his shop what radio station they listened to. The last suggestion backfired once when the people turned out to be three classical and one jazz. Another thing he might do was ask the manager of a shop to turn his radio on to KMPX, if he had an FM, and then ask his customers if that was the station they listened to. When he went into a department store or a place not catering to the hip group, Laughlin would say that despite how he looked, he was there to talk business and to tell them about the new advertising medium available to reach the youth hip group.

Towle was similarly selective about the retailers he approached. He picked ones he thought would appeal to the KMPX audience which, he felt, was a pretty select group of freaks at first and then it broadened out to the weekend hippies. In October, Towle used testimonials in his efforts to convince potential advertisers. He showed them letters from store owners and managers who had tried KMPX and already had good results. The station by then had letters from the Bead Freak, Gramophone Shops, *Ramparts Magazine*, Mainstream Records, and the Town Squire men's clothing store. Sometimes, instead of showing a letter, Towle would make a phone call from one store to another to get a prospective buyer to speak directly with one who had already been helped by the station. He felt he could be honest and straightforward in selling time on KMPX, it was not lies.

In October, the salesmen began visiting some of the larger advertising agencies in the City, building up to do a big run through the agencies just before Christmas. The agencies handled national product accounts and could promise more money than local retailers. As of October 1, Melvin had a list of seven agencies which had advertised on KMPX prior to September 20. All but one were small local agencies. He had a second list of eight agencies he had contacted prior to September 30 who had not placed buys. On a third list, he had 28 agencies he intended to visit starting October 2 and the names of accounts they handled which he felt might do well on the station: airlines, beverages, cars, and movies mainly. He also had a letter from an account executive for the agency handling Levi's who said he had tried KMPX and felt the station had a high percent of influential early adapters, the trend-setters in Levi's San Francisco market. But he had to say he "felt" this, because the station had no ratings yet to speak of, and Levi's, as a national advertiser, ordinarily relied heavily on ratings.

Among the seven agencies who had already placed ads, one was Lennen and Newell where the media supervisor recalled Melvin came to her in purple pants, beads, and a stove pipe hat. She knew what he was talking about because Avery had spoken with her earlier. But other people in her agency did not and some of them were afraid of him. Among the agencies on Melvin's second list was Post, Keyes and Gardner which had Pepsi Cola, Burgermeister beer, Roos Atkins clothing, and Schweppes. The buyer there said she could see at the time KMPX was a coming thing. She knew what was happening in the record stores and among the young. It seemed to her Melvin and Laughlin acted like other salesmen she met. They only looked different. She could see how that got them attention, it got them hearings in places. But KMPX was not much more than amusing at her agency. People were waiting to see the ratings.

Melvin remembered sometimes finding it embarrassing going around to the agencies. He was not sure when, but going into BBD&O with Donahue once, they were walking through a large typing pool on the way to the office of the buyer and he heard all the typewriters stop. Donahue said he felt uncomfortable going around to the agencies, especially at first. He did not want to change anyone's mind. Then he realized he did not have to do that. He was doing the agencies a favor, delivering them an audience and telling them about what would go over with them.

The KMPX salesmen finally got numbers to show for themselves in late November when results of the August-September Pulse audience survey came out. With the numbers now they could say yes, it had been confirmed, KMPX was reaching the 18-34 age group. Its audience was concentrated among 18-24 year old men. A summary used for sales purposes showed the station looked best in the evenings, in the 7 p.m.-midnight time period Monday through Friday, where it was number one in the market for the 18-34 age group. The station had more listeners in that group that either KYA or KFRC, the City's two leading top-40 stations. The Pulse showed an estimated total of 11,900 men and 7,300 women in the 18-34 age group listening to KMPX between 7 p.m. and midnight on weekdays. Of these, 6,600 of the men and 6,100 of the women were in the 18-24 subgroup.

The salesmen had been selling time through the summer and into the fall without the help of ratings, mainly to local retail businesses. For the month of July, records showed a gross billed of approximately $11,000. By November, it was $15,000. About half the amount billed each month was recorded as received. Avery continued to make loans to Crosby. Crosby had notes indicating he received either $30,000 or $50,000 from Avery during July through November. Avery had notes indicating it was $40,000. According to Crosby's notes, Avery issued his loans in $10,000 amounts on July 24, Agusut 27, September 11, and probably on October 5 and November 7.

Johnson continued to engineer for McClay, Donahue, and Voco through the summer and fall months. Of the three of them, she felt Voco was the one who really made her work. It seemed to her most of the people at KMPX were rock'n'roll types. Donahue was, McClay was, and Hamilton carried around in her head a catalogue of names of groups and titles of rock'n'roll songs dating from when she was in high school. But for Voco, the blues were more important. He and Donahue differed about that. Johnson said Donahue would say KMPX played all kinds of music, but he would tell Voco in private he was not going to play blues on his show like Voco did. For him the blues were depressing. But Johnson felt Voco influenced Donahue, however much he denied it. Melvin felt Voco was the one who made KMPX different musically from other stations. In addition to working as a salesman, Melvin sometimes did an air shift which he called The Lone Ranger Show. He said he felt if it had not been for Voco, KMPX might have gone all hits. Donahue was schooled that way and he set the pattern. But Donahue did not feel he was playing hits, he was playing what he liked, which was what he had always done, even in the days of top-40.

By the end of August, Johnson felt she had a good working relationship with Donahue. She could pick up on his cues about what to do next, what to adjust, what record or commercial to play and how to modulate it. They had gotten more and more precise. Sometimes Donahue would take a break, a little vacation, and go out into the office leaving her to progam his show, usually from a list. She would pull the records and play them herself, like Miller did on the all-night show. McClay's engineer, the one he found waitressing at the Minimum Daily Requirement, sometimes programmed his show, as McClay would be tired from working at Tempo and would fall asleep during his shift.

Johnson recalled when she worked with Voco, he would tell her to listen to the last note on something he had just played and then leave it to her to segue in the next cut. Together that way they could take 10 or 15 minutes of guitar solos off different records and make them sound like one piece. Then they might just sit back and say, "Wow." Street said Voco liked having Johnson engineer for him because she had perfect pitch. She could make transitions better than he could. Voco was curious, Johnson thought. He would play rotten music. You would not like it unless you knew what he was doing. He might find a good drummer part on a schlock record, but if you did not know to listen for the drummer part, or to pay attention to how it fit with something else, you would probably think he had bad taste.

Johnson felt Donahue and others at the station were always trying to get her to take acid. They all seemed to be on drugs. If you acted a certain way, they said you had been taking speed. She felt she could not take dope and work at the same time. She drank a lot of beer during the summer and fall. She told them no when they wanted her to take acid, and they would say, "Oh, you mean you are just that way." Owsley, the Acid King, sometimes hung around the station. He was in his early thirties, knew Donahue, had managed the Grateful Dead for a while, and liked to put together sound equipment as well as manufacture acid. He had a radio operator's license. But that was not why he came to the station. When he came he would try to seduce the girls. He would lay little presents on Johnson and the rest of them.

Johnson recalled a journalist for the Washington *Post* came by the station in July in the beginning of a visit to San Francisco to find material for a story on the Haight-Ashbury. He ended up writing a series of 16 articles which appeared in the *Post* in October, under the title, "The Acid Affair." The thirteenth of the series, published October 27, reported that KMPX was the radio station all the heads listened to in San Francisco, and that Donahue, its program director, was a "huge bearded man with a strand of beads, a sinister face, and a tiny old lady named Raechel who rolls his joints and takes care of his correspondence."

Donahue felt the writer of the series, Nicholas von Hoffman, put them all in a bad light by saying as he did that KMPX was a radio station with a format of institutionalized dope music and service to the drug community:

> the only one in the world where you'll see hips, frozen in the lotus position in the lobby. The Haight comes there every evening to chat, to ask for announcements to be put on the air, or listen to the music and use the crayons and paper supplied by the management for itinerant speed freaks who have nothing to do with their hands.

Von Hoffman also said that Donahue expected to put the format into effect on a station in Los Angeles, if the FCC approved. In the Complaints and Compliance Division of the FCC, in late October, a copy of von Hoffman's article went on file in a folder marked KMPX-FM, at the request of the Division Chief.

When von Hoffman initially arrived in San Francisco, one of his first stops had been KMPX. Johnson remembered he had on a clean trenchcoat. Donahue said he talked with von Hoffman about KMPX and what he might find in the Haight, then had him out to his house on Scott Street. He gave him dope, good dope, and von Hoffman got stoned like a baby. Hamilton thought von Hoffman did not initially expect to stay in San Francisco as long as he did. He ended up staying through the summer. He stopped in at the station before he went back to Washington. Johnson said she saw, his trenchcoat was a mess.

Donahue's sister called him from Washington when von Hoffman's October 27 article came out and read it to him and he was disturbed. Johnson thought he may have been afraid of what his father would think if he read about it in Washington. Donahue said no, he was mainly concerned about Crosby's reaction. Von Hoffman's column was syndicated and appeared weekly in the San Francisco *Chronicle*. Donahue said he called the *Chronicle* and asked them not to run the article. They did not run it. The word was the *Chronicle* was going to discontinue the series anyway. Gleason said they felt at the *Chronicle* that the later columns in von Hoffman's series were dispensable, and the *Chronicle* was already covering the subject itself, in a different light.

Not in response to von Hoffman's article, but as stories about KMPX began appearing in the press, Crosby became worried. In the very beginning, press coverage of the station seemed benign. The underground *Sunday Ramparts* carried a piece of straight reporting on it in an issue of April 23. The *Nojo-Navigator*, a San Francisco based rock'n'roll newspaper, mentioned it briefly in April, as did Gleason in an ad lib in his *Chronicle* column. A high school paper featured it in June. Then on July 28, the Berkeley Barb, a paper Crosby did not respect, gave the station its first distinctly whimsical feature-length coverage.

On page six in the *Barb*, a large photograph of Donahue and Miller ran above a four-column article on KMPX. The writer of the article indicated he had visited the station and it seemed to him everyone there was having a good time, but the place looked like a Druid nightmare. A replica of a human brain with veins of various colors sat on one of the studio speakers, changing its color slowly and seeming to pulsate in time with the music. The *Barb* writer said he thought nobody at 50 Green Street believed it, but the station would probably be busted in the next four to six months. The giant hook of the FCC would reach down. Not that the FCC ever listened to what was on the air, but the FCC would respond to pressure and pressure was on its way. Until then, ghostwaves and the midnight cackle of a half-juiced Miller would be playing it like it was. Tao.

On Wednesday, August 16, Gleason devoted one of his entire *Chronicle* columns to KMPX. He felt he was doing the station a favor, giving them his audience. Boucher said yes, Gleason did them a favor. He ignored them for four months. In his column of August 16, Gleason reported that KMPX was the freshest and most important voice in pop music broadcasting since the heyday of KYA. It did community service, made announcements for the Haight-Ashbury Medical Clinic, the Job Co-op, and the Switchboard. The Gray Line hippie bus tour stopped there on its route. He quoted Donahue as saying the station was not only playing music for hippies, but offering an alternative. It was appealing to people who had hip taste but were not into the clothes bit. Gleason concluded that the station had already forced changes in AM radio broadcasting in San Francisco and become a strong factor in record sales. It was throbbing with life-energy and had so many volunteers Donahue did not know what to do with them.

Gleason, in the late summer of 1967, in addition to writing for the *Chronicle*, was involved with the start of the rock music newspaper, *Rolling Stone*. The first issue of the *Stone,* appearing late in October, postdated November 9, included an article by McClay about a radio station in New York that had been doing some programming like that of KMPX. Two weeks later, the second issue of the *Stone* ran an article by Donahue angrily attacking top-40 radio and describing how, for the past six months, KMPX had been experimenting with a new kind of contemporary music format. Donahue's article ran below a half-page picture of the KMPX staff, 26 of them formally posed as a group. Five wore cowboy hats, three held rifles, one held a ball and chain. They were flanked by two dogs. They did not include Crosby or Hunt.

On November 11, *Billboard*, the newsweekly of the recorded music trade, ran an article on KPPC confusing it with KMPX. *Billboard* had first mentioned KMPX in June and had subsequently toyed with names for the KMPX type of format, one of which was "hippop." In December, Billboard seemed to settle on calling the format "progressive rock." Donahue did not like that name anymore than he liked "underground." But by December, much had been done in the name of KMPX and only some of it was to Donahue's liking.

People had been writing to the station all along indicating the curious nature of its audience. On July 9, an architect had written on behalf of Morningstar Ranch, a commune located north of San Francisco near Sebastopol, asking the station to announce that Morningstar was in need of a milking cow for its inhabitants. He hoped one or some of the KMPX listeners would have access to a cow they could donate.

On July 27, a listener wrote from San Jose saying he and many of his friends in the Santa Clara Valley were in need of help. When KMPX had started changing its format in the spring, the word got around quickly and he and his friends became regular listeners. Then a few weeks ago, an evil villain, KPLX, came on the air, broadcasting on 106.5 FM and blocking out KMPX within the Santa Clara Valley. Even in the hills toward Santa Cruz it was difficult to separate the station. The listener asked if one of the KMPX engineers could come up with an antenna trap or filter that would weaken the signal from KPLX. It would mean a fair-sized audience for KMPX rather than no audience at all. Please help, the listener said, all that was needed was the schematic and values for the components.

On September 29, a fire tower lookout on Mount Sanhedrin in Mendocino National Forest heard tabla and guitar pieces played on KMPX at 10:22 p.m., 10:25 p.m., and 10:27 p.m., during Donahue's "Methods of Madness" program. He wrote to the station asking who the pieces were by. Hamilton wrote him back saying they kept no record of what was played during "Madness," but she would guess it had been Alla Rakha on the tabla, however it might have been someone else off an album called "Drums of North and South India." It was Peter Walker on guitar.

Early in September, San Francisco antiwar organizers made known their plans for a series of demonstrations protesting the draft to be held the week of October 16-21. On October 9, Donahue issued a memo to the KMPX air staff:

> Just a reminder that KMPX is a music station. Stay away from political comments or opinions. And since we do not broadcast the news, stay away from it unless it involves music or musicians. The music is sufficient to speak for us.

Chapter 3

KMPX and KPPC
(November 1967-March 1968)

In this period, while the station continues to expand along the lines previously noted, there is an increasing sense on the part of the staff that things are going wrong. Sources of the trouble are identified on personal grounds and persons in positions of greatest authority are assumed to have key roles. What seems to be occurring is an increase in internal demands in attempt to keep pace with the station's apparent external success. There is a buildup of such demands and little meeting of them.

Steven Hirsch believed he was a bear, a bear named Edward. The first week in November, Edward Bear took over the 11 a.m.-4 p.m. shift on KMPX, replacing a disc jockey who had come to KMPX from the top-40 station KFRC in the summer and left in October. Bear said he first heard of KMPX when he arrived in San Francisco in September after driving across the country from New York. He had left New York early in the summer to take a job with a San Francisco magazine, the *Psychedelic Review*, but had been waylaid in Colorado where he spent some time in the company of a chipmunk family.

Johnson was in the studio the night Bear first came in to make an audition tape for Donahue. She remembered he was arrogant, she got angry with him and kicked him out. He came back in a more reserved mood the following night and made a tape which in his recollection had everything on it: rock and classical music merging into Indian and building up to an overwhelming cosmic thunder near the end, like the universe forming and speaking to you. Johnson said she asked Donahue about Bear after he made his tape. Donahue told her he did not especially like Bear but he was going to hire him anyway because he felt Bear had potential.

When Bear sat down to the microphone at KMPX on Monday, November 6 and announced his presence on the air, he had in mind doing a beautiful thing, getting in touch with the souls of his sisters and brothers. He did not believe it was his potential which qualified him for the job, nor was it only his taste and talent in handling music. He was 29. He had dropped out of college in 1962, gone into the Army for seven months, done writing and poetry, hung around Greenwich Village and the East Coast artists' circuit, worked as a bartender, as manager of the Cafe Figaro, and as a classical music disc jockey and program director on a station in New York. He considered himself a man of many abilities and not a radio man primarily.

About the time Bear came on, Melvin and Hunt were recruiting conventional-looking salesmen to help the station get advertising agency accounts. They had run into problems with people who were too hip, Melvin said, people who hung around, boogied, and smoked grass, and thought the money would somehow come in. They took on Whitney Harris in November hoping he would be different. Harris did not have a moustache or a beard. He was bald, 33, and enthusiastic about KMPX. He had come to San Francisco from New York in 1965 with his wife and two children, driving across the country in a Volkswagen camper. He had worked in advertising and publishing in New York. In his last job before KMPX, he had been a sales representative for the United States Chamber of Commerce in Menlo Park. Harris remembered the day KMPX took him on to try him as a salesman, they took on eight others. Two weeks later, he was the only one left.

They tried out six more salesmen in late November and early December. One of that group brought Mr. Broadway onto the station in January. Mr. Broadway was a small hairdresser shop at the corner of Broadway and Van Ness, owned and managed by Stan Weinberger, who had started the business in 1965. Weinberger had not done any radio advertising before KMPX and his shop did not have much of a following. Then in less than three months of advertising with the station, not only did the following of his shop increase, it seemed to be made of more than customers. They were like fans. People associated with the rock groups came in. They asked him if he was Mr. Broadway. The clientele was young, 18-25, and almost exclusively female.

Weinberger said he felt advertising on KMPX totally changed his business. It changed his clientele from straight to hip and made him a semicelebrity. It also changed his attitude about what he was doing. Up until that time, he had not thought hair styling was a trend of the day. It was a business he had gotten into. One of the facts of the business was that people were afraid to wear long hair. But advertising on KMPX encouraged them not to be afraid. Weinberger felt this began with his first commercials in mid-January when the announcing disc jockey said the station's chick engineers went to his shop and that it was groovy and a gas and KMPX was broadcast there. Weinberger put earphones in the hair dryers so his customers could listen while drying. Occasionally one of the KMPX announcers would dedicate a song "to the folks sitting under the dryers at Mr. Broadway's." But the Board of Health objected because of fire hazard and after a few months Weinberger had to take the earphones out.

By the time Weinberger's advertising went on the air in January, KPPC in Pasadena was into its third month of operation as sister station to KMPX. Donahue was program director at KPPC. In his absence at KMPX, Prescott became operations manager. About the same time, Hunt was named general manager of KMPX and Melvin replaced him as sales manager.

Johnson felt things in November began to get out of hand. Donahue's show weekday nights on KMPX was not as regular as it had been. Donahue was now responsible for two shows each weekday night, a new one on KPPC and his old one on KMPX. He would arrange to cover his shift on KMPX when he was down at KPPC, sometimes by taping his show in Pasadena and sending it up to be played in San Francisco the following night. The tape unpredictably arrived late and someone would have to fill in for it. At other times, Donahue would call in a playlist to KMPX before his shift. Someone would take it down. Then doing his shift, Johnson might find the records and run the board while Melvin sat in at the microphone and announced them. On some nights, Donahue's eldest daughter covered for him, pulling the records while Johnson announced. On other nights, Voco filled in. There began to be many mistakes on Donahue's show. Johnson said she heard them but she thought for the most part the audience did not.

Johnson went down to Pasadena several times shortly after Donahue and Hamilton started programming KPPC. They needed help because there were no good engineers down there. The KPPC studio was set up for combo operation and she and Donahue had to sit side-by-side to do his show. One night down there at the home of a friend of his, she took acid. Hamilton was with them. They got someone else to do the show.

By mid-December, Johnson had been working full-time at KMPX for six months and she thought maybe it was getting boring. One of the other chick engineers wanted a raise. In late January, she and the other two full-time chick engineers got together to talk. She was on the North Beach payroll, the other two were on the Crosby-Pacific payroll. Donahue had said he would be willing to raise her hourly rate, but Crosby did not seem to have money for raising theirs, so they did not push it. Johnson felt angry that Crosby was not spending more than he absolutely had to. She had been after him to buy new equipment for the station and he would not do it. It was almost six months since they had gone full-time in the rock music format and the only new equipment they had was from trade-outs. They were still using pieces of home equipment. She thought Crosby should have money to spend on equipment because the station was making it, but she understood he was buying out some of the other owners in Crosby-Pacific, putting money into his wife's house in Marin, subsidizing his girl friend's wig shop in Fremont, maybe buying cars. Much of the talk about Crosby's outside spending was rumor, she knew, but the money had to be going somewhere.

The payroll account of KMPX at the North Beach branch of the Wells Fargo Bank was always low. Crosby had previously had the account at the United California Bank in North Beach but had to move it when he kept getting in trouble. Johnson remembered the account at Wells Fargo ran out on payday most weeks beginning in December. When it did, whoever it ran out on would come and tell Hunt. Hunt would call the bank to find out what the overdraft was. The following Monday, or as soon as he could the following week, he would put in the difference. Hunt said he had to admit the checks bounced. But he always made good on them. When he learned they had bounced he would hustle up enough money to cover them and put it in the account. Then the staff could resubmit their checks. They all got paid in the end. Hunt felt the problem was the station was spending $2 for every $1 that came in. They were always expecting money, but sometimes it did not come in on time. The problem was the money was coming in irregularly. Payday, however, was regular.

The station's bookkeeper felt the pressure. Crosby's accountant who found her said she would cash her check last to save someone else's check from bouncing. Eventually she got an ulcer and had to leave the station. She came to him then and cried. Hamilton noticed how the bookkeeper from Crosby's accountant would make out the checks and then cry when they bounced, as if it were her fault. Donahue felt, whatever happened, people should be paid. They should not have to wait a few days for Hunt to hustle up money to cover their bounced checks. Donahue felt he had always worked at stations with lousy equipment and lousy managers whom he had to placate. It did not seem exorbitant to expect to be paid.

Johnson was switched to the KMPX payroll late in January and her checks started bouncing along with the others. When it happened two weeks in a row, she went to the National Labor Relations Board and reported it. The NLRB said they would collect it for her and then went to Hunt with the complaint, and Hunt came to her. He practically cried, Johnson said, telling her she should not have gone to the NLRB when she knew it was a mistake. He said she had hurt his and Crosby's feelings. He made it into a personal offense.

Johnson and the other chick engineers finally got a raise to $2 an hour in the beginning of February, after the national minimum wage was increased to $1.60 an hour. Johnson understood their raise would come out of commissions of the station's salesmen because they were making increasing amounts each month. The salesmen were working from a new rate card as of January 1, which raised some of their rates by as much as 60%. This was the third advertising rate increase since Donahue began. It was early in February, Johnson thought, that things started getting badly confused at the station. She assumed they got bad because Donahue was not there. He had a way of smoothing things out. Prescott, who was operations manager in his place, did not. Johnson felt Prescott was a little tyrant as operations manager. He was also a tyrant as a disc jockey with his engineers. So was Bear.

Midwinter at KMPX, as Johnson recalled it, was an uncomfortable time. People were taking equipment from the control time. People were getting on each other's nerves. Records kept disappearing from the library. Early in February, Crosby began calling staff meetings, he had not before, although Donahue had called staff meetings when he was there. At first, Donahue held them almost weekly, later only when it seemed necessary. Johnson said she felt Donahue's meetings were not like Crosby's. Crosby's were a series of accusations. In one meeting early in February, Crosby accused her and another chick engineer of coming in late. He did not seem to realize that the other engineer's main reason for working at the station was that she was a groupie. Then he got furious and delivered an ultimatum to the entire staff. He stood up, pointed his finger, went rigid in the face, and said if anyone played dope lyrics, or came late, or swore on the air, he would be fired on the spot.

Prescott remembered Crosby at that meeting raising his finger saying, "I'll fire every one of you."

On February 12, after Crosby's ultimatum, Hunt issued a memo to the staff:

(1) Employees are to cut down on personal phone calls and long-distance phone calls that are unnecessary and only tie up the lines. This also applies to incoming phone calls.
(2) You heard Mr Crosby's statement regarding the playing of records that contain profanity.
(3) No comments or opinions tying in KMPX in any way, such as use of stationery or speaking on behalf of the station regarding your opinions, on any issues to anyone without approval from myself.
(4) Employees are asked to be cautious in their use of swearing and loud noises in the station.

(5) Effective March 1, 1968, employees will be paid every two weeks rather than weekly. Paychecks will continue to be distributed on Friday.
(6) Employees are to be prompt in their working hours. Overtime for DJs and engineers must be approved by Bob Prescott.
(7) Employee benefits (sick leave and health insurance) are presently being studied and considered.
(8) Employees are asked to wash out their coffee cups immediately after use and to empty ashtrays. Also, please clean up immediately any liquid spills.

Hunt believed that by mid-February, Crosby was thinking everyone at the station was against him. He thought even Avery was against him. At first, Hunt said, Crosby had just been suspicious, you know, in that skeptical way he has. Then people began insulting him and he got afraid for the license. Some of the songs they played were profane, like "The Pusher." Donahue said it was actually a condemnation of heroin. Crosby heard that it said "goddamn" eight times. Hunt felt it was not the kind of thing to send into people's homes. There were laws against it. Boucher knew Crosby and Hunt were touchy about lyrics. They had issued a memo forbidding the staff to play certain records. He felt he was helping everyone when he turned off the station's house monitors sometimes to keep them from hearing what went on the air.

At one point, probably late in January, Crosby told Melvin, who was then sales manager, that he wanted the salesmen to take any accounts they could get, as he had to have money to meet the payroll. Melvin recalled feeling horrified. The whole point of selling time on KMPX was to be selective about accounts, to take only what seemed in keeping. Rather than take any accounts as Crosby had requested, he went to a friend of his who was a dope dealer and had savings and asked him for a loan of $1,000. His friend wanted to give the money to KMPX as a donation but Melvin told him no, it had to be a loan. It was a loan, in small bills which Melvin's friend dug up from where he had buried them in his yard. Some of the bills had dried mud on them when they handed them out to the staff on payday. Donahue remembered with something more than amusement the week they met the payroll with "muddy money." When Melvin delivered the money to Hunt to distribute, Hunt asked where he got it. Melvin said he was expecting the question and told Hunt he had an independent income. As far as he knew, Hunt believed him. Hunt and Crosby were not turning things down.

They were grabbing pennies, Melvin said. They could not understand that the station might be a success. In six months to a year, they would have enough money to cover it all. Hunt did not agree. Toward the end of February, he felt he and Crosby were beginning to believe it could work. The station's February sales gross was nearly $25,000. They thought they could see expenses leveling off. They were considering ways to provide the staff with benefits: sick leave, health insurance, vacations. Crosby felt, in his own way, he understood success better than any of them. As you became successful, you got more complaints from sponsors and other people. They want to get you to keep you down.

The tensions were also affecting Donahue. On one of his trips back to KMPX in late January, after becoming aware of discomfort on the part of some of the staff with Miller, Donahue called Miller into his office and went through a list of criticisms of his show. Miller said he thought the criticisms were not accurate. On the following Tuesday just after midnight when he came into the station and started doing his show, Miller found a memo from Donahue reiterating the criticisms: lateness, failure to cross-plug other programs, mistreatment of commercials, drinking in the station, having nonmusic people on as guests, carelessness in keeping logs, inadequate production quality. Copies of the memo had been sent to Crosby, Hunt, and three other administrative staff. Miller got mad, read the memo on the air, and posted it on the main bulletin board in the station. Later in the day, he was fired by Hunt. Crosby was out of town.

On February 15, the San Francisco *Express-Times*, a new underground newspaper, ran Miller's photograph on its cover. Inside on page one was a half-page picture of Miller and an article drawn from an interview he had given an *Express-Times* reporter who had spent an afternoon with him at a Folsom Street bar, drinking steam beer and discussing his dismissal from KMPX. The article said Miller's dismissal had been abrupt, his program on KMPX was chaotically beautiful, Donahue's memo was tersely worded, and this was surely a case of creeping commercialism. Miller did not agree with the last, he felt there were other reasons. He told the reporter Donahue had objected to his talking on the air with people not in the music business. He had Paul Krassner of the *Realist* on before he was fired and he understood Donahue did not like that. He also had some Diggers on.

The chick engineers did a ladies show from 7 p.m.-10 p.m. on Sundays in February. Street said they bitched at Donahue and finally he gave them the show. A man who had engineered foreign language programs before Miller's time, who had been replaced by Prescott in May, returned in mid-February and took Miller's all-night shift. Hunt temporarily fired Bear for playing a long cut of classical music one afternoon, then rehired him apologetically when he found out it was a mistake to take a man off while on the air. Gleason did a ten-minute program three evenings a week during the San Francisco *Chronicle* strike, on the days his column would have appeared in the paper.

At one point in February, Prescott proposed a schedule for switching around the shifts of all the announcers for a week so no one would be in their usual time slot. This would change the routine he felt had settled in and dampened people's spirits. It was a Donahue device to propose switching the shifts, yet Donahue knew what was wrong at the station would not be much changed by it. Like Johnson, Donahue felt things had been getting out of hand, but many of them were ordinary things. Toward the end of February, Donahue sent a memo to Prescott:

> Bob:
> I don't know who's typing the copy for the book, but it is unbelievably rotten and I cannot imagine any reason why it cannot be typed without the unbelievable multiplicity of errors, misspellings, and general fuck-ups all over the copy.
> In addition, Patch's copy is the worst crap I have ever read as long as I have been in radio. Specifically, New Monk Dance and Blushing Peony (which is not only written in a ridiculously "cute" style, but was obviously typed by a one-fingered cretin). I hate to burden you excessively, but I believe we have come to a time when some sort of continuity acceptance is necessary. There is just some stuff here that no matter how many times you read it in advance it will not make sense and there is no way to do it without rewriting it.
> The music library is in disgusting condition. Again this week I have received no list of what went into the library and thus must try to tape my shows against almost insurmountable obstacles.
> Item: Tonight I wanted to play Ultimate Spinach to show where they had gotten some of their material from country, but it was not in the library.
> Item: There is only one copy of Hapsash and the Colored Coat. I told her to order a second copy of this last Wednesday.
> Item: There have been three new Dave Van Ronk LP's issued in the past two weeks and we don't have any of them.
> Tom

The station's gross billings continued to increase through the winter: from $15,000 in November to $24,800 in February. Sales for the first two weeks in March indicated the March gross would be a minimum of $29,000. Crosby's notes showed he continued to receive loans from Avery. Yet expenses continued to increase and Crosby felt pressed. Toward the end of February, he made it clear again to Melvin who thought this time maybe they would turn up something in the agencies.

Melvin and Laughlin then went to Post, Keyes and Gardner, an agency they had visited in the fall where the media buyer had said she recognized KMPX was a coming thing but was not ready to place a buy. This time she placed one for Pepsi Cola to go on the air in the start of March. It was a jingle, a Pepsi generation jingle. The air staff objected. Bear found the spot abhorrent. McClay was beside himself. It was not so much that Pepsi rotted the teeth. This was the first national product jingle KMPX had run. It was prerecorded, upbeat, and slick. It was what advertising agencies were about.

The staff called a meeting to discuss the Pepsi spot soon after it went on the air. There was a hue and cry, Donahue said, led by people like Bear. The air staff wanted to do away with the spot. Melvin proposed they keep it. He felt he and Laughlin had gotten it to play for time with Crosby and Hunt. The air staff still protested, they felt this was selling themselves out. Then Voco stood up and said he drank it, they all did, and they had to agree. But they wanted the spot changed. Someone suggested they ask Pepsi for their old jingle, "Pepsi cola hits the spot." It turned out they could not get it. They offered to produce a commercial for Pepsi themselves and Boucher tailor made one. But it was refused by the agency and he added it to his collection of spots the station never ran.

The media buyer at Post, Keyes and Gardner said she knew with the Pepsi buy she had placed the first jingle broadcast on KMPX. The fellow who played it announced it was "a jingle." She told Laughlin the station could not give that type of treatment and keep the account. Commercials could not be made differently for every station. Advertising was repetition, sameness. KMPX could not win a fight for control of commercial copy. Win or lose, a writer for the *Express-Times* of March 14 suggested it did not matter:

> There used to be this word "hippie." I'm still fond of it, but it doesn't serve like it used to. So I'll call us the Community. In a society that seems to be breaking into Establishment White vs. Black, Lyndon vs. Stokely Carmichael, we are emerging as the Third Force, like a coalition between Yugoslavia and India. What you might call the mostly white alternative. In a word of KFRC vs. KDIA, we are KMPX, complete with all the contradictions of people who advertise Peace & Freedom, Record City, Pepsi Cola and the Highway Patrol on the same station.

But the possible contradiction between different kinds of advertising was not of much concern to Crosby in the winter months. He had other worries, especially some concerning KPPC. On January 16, the board of directors of Crosby-Avery Broadcasting met in Pasadena. Minutes of the meeting kept by Crosby's lawyer showed they discussed the financial situation of KPPC, talked of plans for constructing a new studio in the Old Town section of Pasadena, approved of Donahue's handling of the programming on KPPC, and agreed to pay Avery a salary and residential expense of $1,500 a month as soon as the station turned profitable. Avery held the position of general manager of KPPC as well as president of the board of Crosby-Avery. Donahue was program director of KPPC, for which he was paid, through North Beach, $200 a week plus travel and commissions on sales.

Since November, Donahue and Hamilton had been spending more than half their time down in Padadena changing the format on KPPC to make it like KMPX. Hamilton recalled they spent almost full-time at KPPC in December and began splitting their time between the two stations more evenly in January. When they started at KPPC they did not have time in advance to hire new staff so they kept the people who were there and wrote them out playlists so they would know what to do. In December and January, they took on new staff, hiring some right off the street. They brought down from San Francisco the disc jockey Bear had replaced in November, whom Donahue felt was an experienced professional and expected to be able to rely on. Several of the announcers of KMPX came down once or twice to do guest shows. Melvin came to set up the sales department. Johnson came for engineering. For many of the KMPX staff, the beginning of KPPC was a peak. They had ideas for trading staff and programs between the two stations and for transmitting live shows. There was an excitement. Donahue felt with KPPC there were all these people in Los Angeles hearing it for the first time and loving it. There was this cushion of time before they would start demanding more, expecting more, criticizing the programs and commercials as they were already doing at KMPX.

Crosby's lawyer found an excitement in KPPC similar to that of the KMPX staff. Hunt thought Crosby's lawyer was like a kid at a candy store about the radio stations and that he wanted to manage them. Hunt said he caught Crosby's lawyer usurping his desk at KMPX. He sat there one day and put his feet up on it. Hunt's desk was larger than Crosby's because he handled the business, and from handling the business he knew Crosby was not getting rich off KMPX like people on the staff claimed. Crosby did not take more than two grand a year from the station, Hunt said he knew, because he had to countersign for him. According to rules of the company, Crosby could not himself take money from the radio station. Crosby's accountant figured there was an amount of $24,000 accrued to Leon Crosby in commissions in 1967, but it was not paid him.

Hunt said Crosby was not in any daily business other than KMPX and KPPC in 1967 and 1968. He was not supporting his girl friend's wig shop, she was supporting him. Records kept by Crosby's accountant showed he invested $6,000 in Wigs Unlimited, but that was prior to April 28, 1967. Crosby's accountant agreed with Hunt that Crosby was not engaged in other business at the time. He had trouble enough with the radio stations. He accepted loans from his wife, his sister, and his girl friend. Crosby's accountant could not quite see how a man would live off women like that. But Crosby managed. He helped in the start of things. He helped his girl friend set up her wig shop. Aside from the equity capital, it was said he initially got her three wigs on trade-outs from KMPX.

But many of the KMPX staff preferred to see Crosby's financial dealings in a different light. To them, Crosby was not a marginal businessman managing surprisingly to stay alive, but a seemingly marginal businessman getting away with murder. Whatever the evidence, they felt he was unfaithful to them and there followed from this considerable speculation as to how and why. A common line began with the fact that Crosby never turned up at the station until afternoon, after 1 p.m., often not until after three. The salesmen were making sales, Avery was giving loans, but Crosby never had enough money for the station, and he acted evasively when asked. It was likely, therefore, that he was involved in some other business and that it took place in the mornings, probably out of town. Donahue suggested if it was not wigs, maybe it was fried foods. Crosby's lawyer asked why Crosby was secretive if he had nothing to hide. Crosby's lawyer said he must have had 20 telephone numbers where he thought he might get hold of Crosby, but he could never find him in the mornings, nobody could.

Crosby said in the mornings most of the time he was at home. He could not get up and face it. It was that bad. His girl friend, she supported him for three years. His lawyer, his greasy lawyer tried to tell him what to do and then turned against him. He finally had to tell his lawyer one day, "Harry, I'm afraid you'll become the owner and I'll become the janitor if I'm not careful."

Crosby's lawyer remembered Crosby put it like that. Crosby was a real flake. He was the type who had to ask for money to buy a cup of coffee or pay for bus fare. He carried hardly any with him. He must have lived on a personal budget of $10 a week. He could never pay his bills. Then one day he came in with Avery's money, Avery who was meticulous, ate his fingernails, chain smoked, and was so polite he made it awkward by reaching to open every door. The three of them set up the corporation for KPPC. Crosby's lawyer said he agreed to join as a board member although he knew most lawyers would not get involved with a client that way. But it was such an opportunity. He was fascinated by it. He thought the studio at KPPC would look good enclosed in a glass bubble. He did not anticipate a career in radio and he had no stock in the corporation, he just wanted to do what was best for the corporation. Crosby and Avery kept telling him they wanted him at their meetings. He could feel they did not trust each other and they trusted him less, but they wanted him there. He felt he was bound by his position to be unpopular. He would be the decisive vote if there was a difference of opinion between Crosby and Avery, and however he voted, he would look bad to the man he disagreed with.

In February, Crosby's lawyer said Crosby came to him with a question about where the money was at KPPC. The two of them flew down to Pasadena on February 15, went to the bank and found money missing. They had a record of there having been $33,000 on deposit as of January 16. A financial statement drawn up by the accountant for KPPC showed $20,000 in cash on hand and money in banks as of December 31, down to $13,000 as of January 31. Crosby's lawyer said he and Crosby could not figure out what happened. They could not find anything wrong. It occurred to him maybe too much was being paid for things. There were reports that someone at KPPC had been stealing. Maybe Avery was just okaying Donahue's expenses. Crosby's lawyer said he never did understand the relationship between Donahue and Avery. He never really knew Donahue, but it seemed to him they were going through money like water at both KPPC and KMPX. He was heartsick when he heard in February about the money gone and thought Avery had betrayed them.

He wrote to Avery on February 12 that he and Crosby would be coming down to Pasadena for a board meeting and that they wanted to discuss construction bids for the new studio, the antenna site, and the corporate financial picture. The last, he said, was the one subject they had never gotten straight. When he and Crosby arrived in Pasadena on February 15, in addition to checking at the bank and looking through the books, they arranged to hire a full-time bookkeeper at KPPC. Then they held their meeting with Avery and agreed to a partial reorganization of the administration of KPPC, putting three vice-presidents formally under Avery, the general manager. One would be a music vice-president (Donahue), one a sales vice-president, and the third an unnamed vice-president. The point, Crosby's lawyer said, was to put some restrictions on Avery and what they were doing down there.

In early February, Crosby put about $13,000 toward the purchase of a third radio station in Seattle, Washington. Crosby's lawyer thought Crosby borrowed on the loans Avery made to him for KMPX to do it. Crosby said no, the money he put into the Seattle station came from his sister.

In the end of February, Crosby and his lawyer asked Melvin to come down to KPPC and be station manager. Avery was in the hospital for an intestinal operation. Donahue thought Crosby and his lawyer may have believed bringing Melvin in as manager was a way to get certain things disposed of in Avery's absence. They may have assumed Avery would die and then the station would be theirs. They may have been trying to keep him, Donahue, out of the general management. But that would have been odd because it was hard to imagine they did not know that he and Melvin were close. Hunt said he did think Avery might die when he was in the hospital and then Donahue would try to take control. Crosby suspected Donahue was taking advantage of Avery while he was in the hospital. He said he knew Donahue once went to visit Avery in the hosptial and got in by saying he was Avery's son. Crosby felt Donahue got away with things like that, and that he did it by telling half-truths and by misleading people.

When Melvin went down to KPPC in the end of February, he did an accounting and found that the sales manager at KPPC had been stealing. He told Crosby but he felt Crosby thought what he said was just another play by himself and Donahue to take over the station, and that was too bad because the only play was for success. Crosby's lawyer recalled he knew the sales manager at KPPC had gone through a lot of money, he promised to pay people more than anyone thought. But the main thing was that Donahue wanted to be manager at KPPC and neither he nor Crosby wanted Donahue to have it. There was a struggle for power between Donahue and the sales manager at KPPC while Avery was in the hospital. Crosby's lawyer said finally he and Crosby thought Melvin should be the one to do it. They thought he might be an appeaser.

Donahue remembered January and February as a troubled time. Crosby and his lawyer acted against Avery while he was in the hospital. They brought Melvin down to be station manager at KPPC. Crosby was increasingly uptight and afraid at KMPX. Donahue was traveling back and forth between the two stations, airport to airport, from the basement of the Presbyterian Church in Pasadena to the second floor of the warehouse near the piers in San Francisco. Some weeks he made the trip several times. Building KPPC was not easy for him. Los Angeles was a new situation. He did not have friends there. He had to take people off the streets. He suspected Hunt was playing manager at KMPX in his absence, and he assumed Crosby had money although he would not show or spend it. In the end of February, things got worse. There were incidents concerning the conduct of staff at both stations. Crosby's lawyer tried to get rid of hangers-on and tried to tell the staff how to dress at KPPC. The checks were still bouncing at KMPX.

Donahue felt he had tried. He had tried to act "with" Crosby and Hunt when he could. He had tried to make Crosby feel part of it. But there were some things he could not do, and Crosby repeatedly drew back. Again one day, probably early in March, he came up to San Francisco from Pasadena and went into talk with Crosby and Hunt. He recalled he told them "we" have to do something, but it must have seemed to them he was trying to seize power. In a way he was, but not like they thought. It seemed to him the situation was out of control. He had not been able to delegate authority to anyone but Melvin. He had hoped he would be relieved at KPPC by the disc jockey he had brought down full-time from KMPX, but the man came down and freaked and could not do a thing. Then Crosby and his lawyer tried to move Avery out while he was in the hospital. They took Melvin for station manager at KPPC.

It was probably on Wednesday, March 13 that Donahue went into talk with Crosby alone at KMPX. Crosby said he called Donahue into his office and told him he could keep his position as program director at one of the stations but not the other. He told Donahue he was missing his shows half the time on each station and it was costing too much money, so he should have his pick and take one. Donahue heard, got up, and walked out. He thought it was KPPC Crosby said he could keep as program director. On KMPX he would only be talent. But which station it was was not the matter, or it was not what Donahue remembered. He thought it was KPPC.

Crosby watched Donahue walk out of his office. He did not fire him, he let him go. Donahue had to be bride and groom, he thought. Donahue wanted both stations. He wanted everything. Crosby had told him to take only one. Crosby suspected Donahue was in league with Avery. He suspected Donahue had all the people on his side, while he, Leon Crosby, was low man on the totem pole and, he could tell, they were trying to push him off.

Chapter 4

THE KMPX STRIKE
(March-May 1968)

In a final demonstration of dissatisfaction, members of the staff of the station go on strike. They walk out of the station and remain off the air for eight weeks. What is remarkable here is that much of the pattern of support for the station on the air continues with the staff on strike. This is shown in benefits and meetings given and attended by advertisers, listeners, and musicians. The main difference in the pattern is an increase in the amount of publicity accorded the station and a split between the striking staff and the station's ownership. There is, finally, the finding of a new opportunity with a larger corporate owner.

March

Donahue ignored Crosby's offer that he take one station. He felt arbitrarily relieved of his duties. Whether he had brought it on or not, he did not care. Crosby had just taken back what was not his to reclaim. The world for a moment collapsed. Donahue walked out of the office and left 50 Green Street. A while later he talked with Melvin. They decided to go on strike. They did not think of going to another station. They did not want to kill this one. They wanted to keep it from being destroyed. You do not kill what you love, Donahue said.

Before his meeting with Crosby on Wednesday, Donahue recalled thinking he and Melvin were fighting to keep the stations together until Avery could get on his feet. Avery was part owner. It was now no longer a question of that. On Thursday, March 14, Melvin resigned his position as station manager of KPPC and word got out that he and Donahue had left. At 50 Green Street, Crosby and Hunt tried to assure the remaining staff they wanted to keep them on.

The weekend before, late Saturday night and early Sunday morning, March 9 and 10, Johnson and Donahue's eldest daughter had stayed up and dubbed about 45 small tapes in the KMPX record library. These were the station's collection of original recordings by Dylan, Joplin, Kaukonen, and other local musicians. Toward the end of the week, after Donahue and Melvin had left, the station's record librarian, Lynn Hughes, stayed up two nights from midnight to 6 a.m. and read into a tape recorder names of the titles and distributors of all the records in the

71

library. She recalled there were between three and four thousand. According to a Federal regulation, they were property of the station owner.

On Thursday, March 14, Tony Bigg was called to replace Donahue on the 8-midnight shift on KMPX. Bigg had left his job at the top-40 station KYA in February after announcing he was quitting on the air. By the time he was called to replace Donahue, he had already done some fill-in work on KMPX and a newspaper story had come out spelling his name with a P instead of a B. He accepted the change and thereafter went by the name Tony Pigg on KMPX.

On Thursday, March 14, Prescott replaced Donahue as program director at KMPX. Prescott understood that Crosby had at first told Donahue he would be relieved of his duties at one of the stations and later, on Thursday, had fired him. Prescott said he was then asked by Crosby and Hunt to take Donahue's place as program director. He had doubts but agreed to do it. He wanted to keep the station from going into instant crumble. He wanted to talk to Donahue at the time but could not. He thought Donahue must have gone home after he walked out of Crosby's office, done whatever he did to recover, and then decided with Melvin to leave.

Friday, March 15 was payday at KMPX. Prescott said everybody knew on Friday that Donahue and Melvin had left. Donahue came into the station to pick up his things. People were walking around stunned. They talked about having a meeting Saturday. Many of them thought Donahue had been fired. When he walked out of Crosby's office on Wednesday, Donahue had shaken his head and said something Johnson thought could only mean he was fired. He may have felt he was fired. He felt he was forced to resign. Crosby said he quit. Crosby said Donahue did not say yes or no about taking one of the stations. He did not name the one he wanted. He just walked out.

On Friday, March 15, Johnson set about to quit her job as an engineer at KMPX. She felt she had had enough. She was tired of coming into the station and getting yelled at. She wrote a letter to Crosby saying that because he had fired Donahue, she could not work there anymore. On her way to mail the letter at the Rincon Annex post office in San Francisco Friday night, so it would have a Friday postmark and she would not have to work that night, she stopped off at Donahue's house and told him. After mailing the letter, she stopped by the station. Someone else was doing her shift. The next morning, Donahue called her at home and told her there was going to be a meeting at Prescott's that afternoon.

At midday on Saturday, March 16, 29 members of the staff of KMPX met in the living room of Prescott's apartment on Greenwich Steps, up

Telegraph Hill behind 50 Green Street. Donahue proposed they go on strike. Crosby had made their situation at KMPX unworkable. Melvin concurred. The staff discussed it and voted unanimously to go on strike. They would let their troubles out. They agreed to a starting time of 3 a.m. on Monday, March 18.

There was debate about the starting time. Some thought they should take the station off the air at peak audience time Sunday night. Others thought they should play it to the end at three o'clock Monday morning, when the transmitter would ordinarily be turned off for three hours so engineering maintenance work could be done. One salesman voted against the 3 a.m. starting time but in the end the staff agreed to convey a buildup to the audience Sunday night. In the early hours Monday morning, they would give details of their planned strike. At three o'clock they would split.

Among the 29 people present in Prescott's living room, there were doubts about the desirability of taking the station off and walking out on strike. But there was a general will to do it and an expectation that their absence would be brief, they would not be out for more than week, maybe for only a few days. They would demonstrate their value to Crosby. He would realize he could not stand their loss and would take them back and stop trying to restrict their conduct. Prescott recalled he thought the staff could intimidate Crosby, that the logic of making money would get to him. Prescott felt the initial idealism they had at KMPX had turned into a working situation by the time the staff met in his living room. Their voting to go on strike was a way of reaffirming that idealism, a way of saying the whole thing was more important then anyone's job, and that it was necessary it not fail. They still had a long way to go, and what they were about to do could not legally be called a strike. It was only a walkout. The staff were not members of any trade union. They formed a union of their own for purposes of the strike: the AAFIFMWW, the Amalgamated American Federation of International FM Workers of the World, Ltd., North Beach Local No. 1. The FM could stand for frequency modulation or free men.

On Friday, March 15, the Friday after Donahue left and before the staff voted to strike, Hunt had accompanied some of the staff to the bank to make sure their checks would not bounce. He went for a beer with Crosby afterwards and remembered saying to Crosby for the first time he could see the way to their making a buck. The financial situation was actually getting better. Expenses were leveling off, the rent was covered, the ratings were good. Then Sunday night, the strike hit. Donahue wanted to take it over. Hunt said he could see Donahue had been frustrated during the past few months because of power struggles. If you are a Hitler you are frustrated if you cannot take over the country. Hunt went to bed at home in Pleasant Hill at 11 p.m. on Sunday night. Crosby phoned him after 12 p.m. and said he had been listening to the station and there was something going on down there. Hunt thought Crosby was not too concerned, that Crosby felt it was a lot of crap, but he did not like it going on the air and he asked Hunt if he would go look into it.

Hunt got out of bed, dressed, and drove 26 miles from his home in Pleasant Hill to the station at 50 Green Street. When he got there he found bands out front, a crowd of people, signs, pickets, a light show, the Dead were playing. He was flabbergasted. He went inside. He came out. He talked with Donahue. They hashed over some of the problems. He did not call Crosby for several hours. He did not want him to find out more. He knew Crosby had a hearing to attend in court at nine the next morning, a lawsuit filed against him by one of the Portuguese program producers who had been on KMPX before the format change and had a political dispute with another of the Portuguese producers.

Hunt stayed at 50 Green Street and talked with the strikers outside. He read their list of demands and thought they were fruitless. They would let Crosby retain ownership, he could keep his managership, and the employees would run the place. Hunt said he agreed with the strikers that the station was the people. However, it was property and owned. The owner was responsible to the FCC, he had to run the station. Hunt felt the staff had been insulting to Crosby for months before the strike. They had tried to keep him out and had not given him a say. It was true Crosby blew up at the wrong times and did not handle it well, but he was the owner. Hunt stayed at 50 Green Street and continued to talk with the strikers in the early hours of the morning. He tried to get a few of them to go inside and go back on the air but they refused. Later in the morning, Crosby came over. He and Hunt went for coffee. Crosby then left for his hearing but he got the feeling, despite what they said, the strikers were not going to negotiate a bit. They had played their hand and that was all they were going to do. They even asked him to go on strike with them.

When Crosby got back from his hearing, he called a labor law firm. The labor lawyers advised him not to talk to the press. They told him something in the National Labor Relations Act forbid him from interfering with a strike and that publicizing his side would be doing that. Hunt said he thought the labor lawyers had in mind a normal factory strike where the workers did not try to kick out the management and take over the plant, they just wanted better conditions. He told the labor lawyers he thought there was a difference, but they could not see it. They advised against publicity.

When Crosby came back from his court hearing and sat down with Hunt on Monday, Hunt thought Crosby was just plain mad. He was not realizing the future. He simply wanted to fire them. Hunt said he told Crosby he thought they should talk to the strikers, for the sake of the station's image. Crosby agreed. He said they would try to negotiate the reasonable demands. Some of the money ones would have to be met, but not the others.

Johnson was a member of the negotiating group representing the strikers which talked informally with Crosby and Hunt on Monday. She had not wanted to go on strike at first because she had already quit. She told the others Saturday at Prescott's house that she did not want to go, but they had her join them. They said they were all a family and they made her a member of their negotiating group. Shortly before midnight on Monday, the negotiating group had an informal meeting with Crosby and Hunt at the station. Johnson remembered Hunt and Crosby kept saying, "You can't do this to us, you can't do this to us."

Hunt said he and Crosby thought Donahue's walking out the previous week was probably done on purpose to bring on the strike.

The KMPX staff had announced their plans to strike on the air on KMPX on Voco's program between midnight and 3 a.m. Monday, March 18. Bear had made the initial statement:

Bear: As you know, more or less, we've been talking of splitting and leaving, and that is indeed what we're doing. If you haven't picked that up directly, the entire staff, and that means just about the entire staff from secretarial help on through everybody, is going out on strike at three o'clock this morning when we normally go off the air. And the reason we're going out on strike is not because there's so much black plague running around but that it seems kind of apparent to all of us that the ability to have creative growth here has been so limited to the point that the foreseeable future is grim.

This does not mean the present is that awful. You have, we have all been enjoying a groovy thing, I think, for quite some time. But the management and the creative staff are truly at odds, simple as that. And I'm sure most people know that Tom Donahue, more responsible than anybody for the existence of KMPX, I expect, quit last week. The entire strike has to do with that. It would be nonsense to say that it didn't. This is in sympathy with Tom. With the same conditions that he had to quit, well, so have we all.

And I really got mixed feelings. My head is in a strange place because I really feel sad, and at the same time rather hopeful that the reason that we're doing this all is so that there can be creative freedom, so that there can be truth, be it in the music or what we say or whatever in broadcasting. What a good thing in communications that it can be truthful.

At the moment we're getting crimped and I hope you don't, I hope you don't feel abandoned, for a lot of people who have spent a lot of time with us. Because that's not the whole thing. Anyway, we wanted to let you know we are going out on strike and we have a bunch of things, I don't know if I should — We're called, by the way, the Amalgamated American Federation of International FM Workers of the World, Ltd., North Beach Local No. 1. That's us. We have at the moment about 30 members and we all work here at KMPX, until three o'clock, and then we're out on strike. We have 11 things that we want considered, and say I'm not to the point of airing them at the moment, but they're things to allow for that freedom.

Yeah, it's crazy. Anyway, so we're going out on strike, at three o'clock. We're going to have, I don't know, did the Dead say they're definitely coming, Voco?

Voco: Right.

Bear: Anybody else?

Voco: The Dead will be here, Almond Joy will be here.

Bear: Anyway, there's going to be a dance.

Voco: Ther cops are going to crash.

Bear: It's going to be chilly outside. If somebody would like to bring some barrels or some firewood, that'll really be nice, and we could have some fires out there maybe. I don't think we're going to get hassled from the heat because they're aware that something's going to happen. There have been announcements in the Fillmore and the Carousel and the Avalon, for people if they want, if they care about

The KMPX Strike (March-May 1968)

	the station, if they want to, anything from be curious to come for support, hang out at about three o'clock. There will be music. There will be us all picketing. And like that.
Music:	We shall overcome. . . .
Hammond:	Good evening. This is Phil Hammond. The moon is in Scorpio until 10:21 p.m. Monday night. The great sphinx looks with scorn upon our station. He has watched while our motives were murdered and our talent wasted. To his mighty disgust, white sheep have been sacrificed to our sponsors with averted face, and hordes of black scorpions overrun our controls. Tonight Pluto, the god of death, dripping crimons stands atop our broadcast tower to signal our redemption. With a face on either side of his head, he looks on the disappointments of our past and urges us to command our own future. The great sphinx stirs at his coming. He knows death is near and his giant wings, unused for centuries, begin to throb and tingle with anticipation. Soon he will fly in the sunlight.
Music:	. . . .
Voco:	Well I'll tell you what. Let me talk over this record a little bit. And we can put it back. Now I got to give ID. We're not that gone. This is KMPX stereo, San Francisco, at 107. This is Voco. And Dusty Street. Ed Bear's here, until three this morning. That's it. We're hanging it up, kind of. If you want to come down and join us here, come on over and say hello and we'll have some music and walking in the streets for you, three o'clock this morning.
Music:	. . . .
Voco:	This is Voco and Dusty until three in the morning and then we split. If you don't believe we're leaving, you can count the days we're gone. Ooh! I must have stole that from B. B. King, long time ago. We're, the whole staff here at KMPX, is splitting. That's it, we're walking out at three o'clock this morning. And we're going to have a little picket party out front. We're going to do a little rockin'. Bring down some water, I'll even stomp on that for you. We're going to have some dancing out there. Listen, KMPX will probably still be on the air, maybe Buffalo Bill

will be one of the announcers, I don't know. Don't help me out. Help me in.

[Music, McClay, Voco]

McClay: This is Bob McClay and I'm up way past my bed time. Only a very heavy moment would bring me in at a time like this. We think we should be clear, and by "we" I mean the employees of KMPX, that we are going on strike at three o'clock this morning. And I think that comprises about, well, an overwhelming majority of the, well, I don't know if there are any exceptions or not. But at present as far as we know there aren't, and that encompasses every employee of KMPX.

Voco: Amen. Amen.

McClay: And we want to make one thing clear. That is that we feel that KMPX has been, up to this very moment and for all intents as far as we know, up until about three o'clock, probably the best radio station that any of us have had the pleasure of working for. And we feel that a lot of good people had a lot to do with putting that together. The reason for the strike is not because of that, of course. It's because we want to keep KMPX as it has been, and this is the only way we see that it's going to be possible to do this, because otherwise in a matter of a short space of time, KMPX will not be the same radio station that you have come to like to listen to. So, with that much rapping, Edward Bear would like to throw something in.

[Bear, McClay, Bear, McClay, Music, Voco, Music]

Dusty: Hello. This is Dusty Street and for those of you who don't know, I'm an engineer here. I usually engineer from six in the morning until noon, but I get out of it as much as I possibly can. I sandwiched myself between two records that I really dig, one of them "Feel So Bad" by Little Milton and the other one I'm going to play, well, I'll let you see what it is. I just wanted to say if it weren't for KMPX, I would never have become an engineer because chicks just can't make it in this business, and it's really a far out business to be in. I really love the people I work for and it's just too bad that we haven't been able to get along like we should have been able to get along. But, like you know, something very, very groovy has got to come out of what we're doing. I'm just very happy that all those people out there are digging us as much as we're digging them.

[Voco, Dusty, Music]

Voco: Hey, make it here at KMPX, 50 Green Street. There's a light show going on downstairs. The people, the groovy people from The Family Dog, the Avalon Ballroom, brought their light show. The Ace of Cups are here, missing one, one ace. Denise, if you're nearabouts and you'd like to fall on by, the girls are here and they'd dig to play. The Dead are going to be here and Blue Cheer is going to be here. Yeah. They're going to rush the Ice House. We're going to have crushed ice next door, baby, this is Voco, and if you want to hear me anymore it won't be here on KMPX for a while. But I'll be on KPFA in Berkeley tomorrow night at midnight on the King Biscuit West, I think that's the name of the show. I'll be there at midnight so, I gotta play the blues somewhere, baby.

[Music, Voco, Music, Voco, Music, Voco]

McClay: They're here! They just walked in with their cannons.

Voco: Beautiful, man, come on in. Play it. Play James Brown.

Voice: This is radio free San Francisco.

Voco: Hey, that's about it folks. We're going to split downstairs. The Grateful Dead are here. Ace of Cups are here. The light show from The Family Dog, we're all here. Hurrah!

Voice: Hurrah!

Indian War whoop.

Voco: Dusty, I got you a gig tomorrow at Mr. Bimbo's, in the fishtank.

Hammond: The time has come again for KMPX to take a brief rest while our team of technical tinkerers tear away at our tiny transistors for the next six hours. KMPX broadcasts on an assigned carrier frequency of 106.9 megacycles with 80,000 watts of power authorized by the Federal Communications

	Commission. We will return to the air at 6 a.m. with a reading from the I-Ching.
Music:	I shake my head and walk away, walk away...Sometimes I wanna stay here, then again I wanna leave...I just can't make my mind up, I shake my head and walk away....
Voice:	Okay. This is now radio free San Francisco. Everybody is free to do as they please.
Music:	I just can't make my mind up, I shake my head and walk away.

Hammering Noises.

Voice: Let's go.

Sound of record playing in grooves.

Sound of a switch flicked.

By the time the last of the **KMPX** staff walked downstairs and out of the building at 50 Green Street shortly after 3 a.m., there were about 500 people assembled in the street outside. Some were listeners who had heard about the walkout on the air, some were people in the music business who had gotten word before. Some, like the bands, had been asked specifically to come. Creedance Clearwater Revival started playing at 3:05 a.m. Blue Cheer, the group Voco produced for Mercury, was there. The Grateful Dead set up their instruments and played. A group of people associated with the Dead wanted to take over the transmitter and free the airwaves. They tried to get Boucher to let them but he wanted to leave the station operational and did. A group from The Family Dog had a light show going. People were dancing near the pickets walking in the street. Green Street was not a wide street. It was shouldered by warehouse buildings. Looking up, one saw Coit Tower on Telegraph Hill. The bands played loud and the music carried up. Residents up the hill complained about the noise and after about 20 minutes police arrived and ordered the crowd to disperse. They tried to relocate at Pier 10 on the Embarcadero but failed to. Someone forgot to carry over the amplifiers. By 4 a.m., the bulk of the crowd had gone. Pickets remained.

The *Chronicle* reported on the walkout briefly on March 18 and said KPPC in Pasadena was out on strike in sympathy. The *AP* ran an item on it the same day. On Tuesday, March 19, the *Chronicle* ran a feature describing the scene outside the station Monday morning which began by mentioning that Owsley, the Acid King, had been there and that while the bands played, joints were passed around, acid taken, the wealth shared. On March 20, *Variety* in Hollywood reported that the staff of KPPC had gone on strike in Pasadena and that KMPX in San Francisco had joined in sympathy.

Once off the air, the striking staff of KMPX started printing up announcements about their activities. In an office on the sixth floor of the Columbus Towers building at Columbus and Kearny, over the hill from 50 Green Street, McClay had a duplicating machine and paper he used in turning out the Tempo newsletter. He donated these to the KMPX group for running off strike literature. The first widely circulated piece they ran off was a notice listing 11 demands the striking staff was making of Crosby-Pacific. The first was that Crosby's lawyer have no authority over supervision of employees, the second that Donahue and Melvin be reinstated, the third that Donahue have complete control of programming, the fourth that Melvin have complete control of sales. The fifth demand was that Boucher have complete control of engineering, the sixth that Blue have complete control of traffic, the seventh that employees share in the increase in profits of the radio station, the eighth that employees be paid in full for time on strike, the ninth that there be wage increases to be agreed upon. The tenth demand was that no employee be discriminated against for reasons of union activity, and the eleventh that all the conditions herein named be agreed to in writing before the strike be considered settled.

Hunt remembered putting KMPX back on the air at 5 p.m. or 6 p.m. Monday evening. He did not worry about getting it back on earlier because he thought the strike would be over by the end of the day. The KMPX strike, begun on Monday, March 18, lasted until Monday, May 13 — eight weeks. The striking group maintained a picket line in front of 50 Green Street the entire time. Volunteers joined them. Crosby and Hunt recruited new staff to put the station back on the air and keep it going. Advertisers withdrew their accounts from the station. Rock groups requested their music not be played. Local bands appeared in concerts to benefit the striking staff. There were formal negotiation meetings between Crosby and Hunt and the strikers. There was publicity, press coverage, and good weather. KPPC was also out. The board of directors of Crosby-Avery Broadcasting met and voted among themselves. Five weeks into the strike, on April 18, a large negotiation meeting was held in the New Committee Theater, chaired by Bill Graham and attended by several hundred people. Crosby spoke at that meeting and said he had been a hippie before any of them were hippies and it seemed apparent he would not take back the striking staff. Donahue and Melvin traveled across the country in search of another station for the staff to move to. On May 21, a group of them moved to KSAN, a Metromedia station with studios in downtown San Francisco. Their strike on KMPX was never settled.

By the time the strike was over, Crosby felt irreparably damaged. Hunt felt the station was dead and that the strikers had killed it. Avery was voted off the Crosby-Avery board. McClay thought maybe it was the first time in history people managed to shut down a business that had to operate. He believed the strike was a success in that respect but it failed because the goal was to go back. The staff for the most part felt defeated, especially when, on May 21, they walked into KSAN on Sutter Street, their new home, and the first thing they saw was Metromedia wallpaper. Johnson drove away from the strike line during the sixth week and did not come back. She thought the striking group was doing horrible things they were not conscious of. She thought the others held it against her for pulling out, and she thought they all changed during the strike but none of them would admit it.

On Wednesday, March 20 at 11 a.m., members of the striking staff of KMPX held a press conference on the ferryboat, San Leandro, tied up by the Ferry Building at Mission and Embarcadero. The strikers had run off several notices for distribution at the press conference and on the picket line in front of 50 Green Street with AAFIFMWW appearing as letterhead on each. The main notice was one which said the strikers of KMPX and KPPC were a "tribe of people" who had brought the new music to San Francisco and Los Angeles. "When these two stations were teetering on the brink of collapse," the notice said, "management had been content to have long-haired, barefoot, and beaded employess." But when the stations became successful, management had seen fit to remove some of the people who created them. Donahue and Melvin had been forced to resign. They then joined the staff on strike. "We love these stations," the notice read, "not as a collection of chairs, desks, tubes and turntables, but as the living idea of a loving group of people. We love our work and only wish to be allowed to do it as we have in the past. For the present, KMPX and KPPC are on the air — the idea is on the street."

On Wednesday, March 20 in the evening, The Family Dog held a benefit concert for the KMPX strikers in the Avalon Ballroom. Eight local bands performed: the Grateful Dead, Blue Cheer, Jeremy Steig and the Satyrs, Kaleidoscope, Charlie Musselwhite Southside Sound System,

Santana Blues Band, Fruminous Bandersnatch, and The Clover. Reportedly, $2,400 was raised for the strikers.

On Friday, March 22, the strikers held a press conference at the Avalon. According to a release they distributed, several recording artists and groups had expressed their support by requesting the KMPX not play their records on the air during the strike. Among them were the Grateful Dead, Blue Cheer, Mimi Farina, and Joan Baez. Jerry Garcia of the Dead was said to have walked into the station at 50 Green Street and asked for a return of the Dead's latest single. Country Joe and the Fish and The Jefferson Airplane subseuqently asked that their records not be aired. The Rolling Stones sent a cable from England on March 21: "We want you to know that we support your battle against the bureaucracy. We believe in KMPX and KPPC and will keep the faith over here. Love, Mick, Keith, Brian, Charlie, Bill." The Peace and Freedom Party voted support of the strike in a statewide convention the same week, saying KMPX had been the Party's nearest radio equivalent.

On Saturday and Sunday, March 23 and 24, a street fair was held in a parking lot near 50 Green Street. The fair was originally planned for in front of 50 Green Street but the site was changed when the San Francisco police refused to grant the strikers a permit for closing off the street. Nine Bands were scheduled to appear. Bill Graham sent over food each day. The New Orleans House sent ham on Saturday.

Within 24 hours after the strike began, all but a few of the advertisers who had accounts with the station withdrew them. Starting the morning of the day the strike broke, the KMPX salesmen now on strike telephoned and went door-to-door speaking with the advertisers, explaining their situation, saying they did not expect to be off for more than a week and felt they needed the advertisers also to leave in order for the strike to succeed. They requested that the advertisers stay off the air until the strike was settled. Thirty of the advertisers formed a committee in support of the strike. On Sunday, March 24, the advertisers' committee held a dinner for the striking staff, their families, the advertisers, and the press. It was reported to have raised $1,000 for the strikers.

Laughlin was in charge of maintaining the strike line in front of 50 Green Street seven days a week, 24 hours a day. The first week the weather was good, it did not rain. The nights were San Francisco spring nights, foggy and chill. Laughlin said he and Hughes, his old lady, scheduled the striking staff and volunteers they recruited to walk the line, day and night, in shifts. They parked Laughlin's Chevy van by the side of the street and organized from there. They served food from the van, monitored the line from the van, and distributed pamphlets and benefit checks from it. On occasion the van was used as a whorehouse.

Shortly after the start of the strike, Hughes spent several days in Voco's Mercury office transcribing the list of record titles and distributors she had read into the tape recorder in the KMPX library a few days before the strike was called. With the list written up, she could send away to record companies and stock a library anywhere just like the one they had at KMPX. She felt it was shitwork transcribing the list, but it was for revolutionary radio and that made it bearable.

By 6 p.m. the first day of the strike, although Hunt had the station back on the air, there was a problem in finding staff. Hunt knew Crosby had run the station before as a radio school. A day or two after the strike began, he and Crosby went out to local colleges: San Francisco State, the College of San Mateo, and one school in Fremont, and recruited students from broadcasting classes, telling them here was an opportunity to get professional experience.

When the striking staff found out that Hunt and Crosby had the station back on with the help of students from broadcasting schools, they were furious. Several of them went out to the colleges. They went, Melvin said, to explain their side of the situation. The students might not have realized there was a strike on or what it meant. They tore down signs Crosby and Hunt had posted on bulletin boards and told students in broadcasting classes if they came to work at KMPX, they would be treated as scabs. Laughlin told a reporter from the San Francisco State *Daily Gator* that he and Johnson and some other strikers had driven down to the College of San Mateo on Thursday, March 21 and literally blitzed the place. The radio and television department was shipping its kids out to the station and he felt they had to stop them. The whole point of the strike was to prove to Crosby that the original staff was invaluable. The scabs coming on proved they were replaceable. Melvin said when the scabs came on that was the first time that thought had struck him. Then the scabs claimed the music was more important than a certain bunch of people's jobs. They said Donahue and the strikers were on an ego trip, that the strike was a power play, that they, the scabs, were keeping the station alive.

Hunt thought in the first week of the strike he must have gone through 150 disc jockeys. He went through disc jockeys like they were going out of style. Some he recruited would not come in when they got to the station and had to face the strikers out front. Some would come once and not return. Some were really bad on the air. For the first two weeks after the strike began, Hunt thought he stayed at the station just about 24 hours a day. Crosby remembered Hunt befriended him. Hunt slept on the floor there in the building with him when the strikers made it so bad he did not dare leave. Hunt said he felt everything seemed to fall to him as the strike continued. He had to take care of Crosby, because Crosby was not up to it and had problems with his girl friend at the time. He had to get money to pay the day-to-day jocks. He had to stall off the long-term payments. The advertisers had gone off because of the strikers' threats. The strikers had even gotten to the factoring service and asked them not to collect. He finally sold time to an Italian program again.

Johnson recalled for the first few weeks, the strike seemed a positive thing. They had really brought radio out to the people. As a member of the strikers' negotiating group, along with McClay and Towle, she met several times in formal sessions with Crosby and Hunt and their respective lawyers. The first formal negotiation meeting was on Friday, March 22 at the Del Webb Towne House on Market Street, the second was the following Tuesday in an office in the Columbus Towers building. It seemed to Johnson that Crosby and Hunt were afraid at those meetings. The group representing the strikers was self-righteous. The lawyers talked. No agreement on demands was reached.

The strikers continued to meet among themselves. Donahue said he decided in the beginning he would not be their leader, it would not be his strike. He would be just a participant. He and Melvin were in Pasadena when the strike was announced on March 18, but they did not tell Avery about it beforehand. What they did beforehand and in the start was to try to anticipate issues. Donahue said he expected the press would be suspicious and the underground press might cause trouble, so they sent them money figures on what the strikers earned from each of the benefits given for them. Various of the staff told reporters what they made in salary while working at KMPX. They said their scale was low, the highest-paid announcer among them, Bear, made $125 a week. They said their checks had bounced while the station was making increasing amounts and they cited those amounts in terms of monthly sales gross. Crosby, they said, was raking it in, not them, they were making more off benefits now than they had at work.

Donahue recalled thinking the demands the striking staff made of Crosby-Pacific had an outrageous aura to them, and he felt Gleason in his *Chronicle* column took the strike more seriously than they intended. Gleason described it as a case of workers seizing the tools. In his column of March 20, Gleason called it the first hippie strike and said it was different, the only one of its kind he knew, ushered in with an astrological forecast full of "gorgeously ominous references to black scorpions and destiny." He said he felt it was inevitable, due to a conflict between the

KMPX programming concept and the station's ownership. "It was a question of style: bearded and beaded announcers, pretty girl engineers, salesmen wearing buckskin and fringe and boots versus the Old-Fashioned Way." Herb Caen mentioned the strike briefly in his *Chronicle* columns of March 19, March 25, and April 3.

The *Barb* of March 22 said the strike was a walkout, in character with what KMPX had been, and beautiful. The *Barb* reported having asked Hunt whether 95% of the station's advertisers had cancelled their accounts within 24 hours as the strikers claimed. Hunt said he did not want to answer and give the strikers the satisfaction. He also declined to comment to a reporter for the San Francisco *Examiner*, saying he wanted to negotiate with the strikers not with the world.

The *Express-Times* of March 21 ran a list of the KMPX strikers' demands in a black box in the middle of an article full of nostalgia for KMPX., The article referred to changes in the station's programming during the 11 months since Donahue began and said that the station had started catching commercials, then commercial commercials, but its basic tone had remained:

> Always its tone was the one in which we most deeply believed business should be done: taking it easy. KMPX's trademark: the programmer or someone fucking-up, blowing something matter-of-factly up front on the air, backed by a chorus of giggles from the bird engineers. I remember, after one song an announcement: "It's okay, Rusty, wherever you are, it's not coming in on the plane tonight, you can relax. This is a public service." Silence. Then a shocked voice from the background: "But you can't say that on the air!" "Well, guess I just did," and Procol Harum flicked on.

On Saturday, March 23 at noon, while bands played in front of 50 Green Street, the board of directors of Crosby-Avery Broadcasting met in the office of Crosby's lawyer on Post Street in downtown San Francisco. Crosby, Avery, and Crosby's lawyer were present. They reviewed a list of 10 demands submitted by the strikers of KPPC, commenting and voting on each. The demands of the strikers of KPPC were, with two exceptions, the same as the demands of the strikers of KMPX. According to minutes taken by Crosby's lawyer and later initialed by each of the members of the board, on the KPPC strikers' first demand, that Crosby's lawyer have no authority over or supervision of station employees, Crosby's lawyer refrained. Avery said the strikers should have no voice in the selection of corporate, managerial, or executive personnel. Crosby said the strikers should have no voice in the selection of directors or attorneys.

On the strikers' second demand, that Donahue and Melvin be reinstated, and the third, that Donahue have complete control of programming, Avery said both should be reinstated with limitation on their powers. He felt it would be harmful to have Donahue as competition and Donahue was knowledgeable about the music. Limitation of Donahue's powers should include the requirement that hiring be discussed with management. Crosby's lawyer said Donahue had ability but was uncontrollable and therefore more trouble than he was worth. He was also untrustworthy. Crosby said programwise and as a disc jockey, Donahue's absenteeism at both stations made him unable to do the work. He was completely incapable of being program director because of his inability to control disc jockeys, his insistence on playing objectionable records after repeated warnings, and his basic dishonesty, an example of which was the *Newsweek* article of March 4 which reported that Donahue was hired by Crosby to manage both stations when he was not. Crosby said Donahue was also at fault for his attempts to control all station personnel and his refusal to restrict himself to programming. An example of the last had occurred when Avery was in the hospital. Under no conditions, Crosby said, was Donahue to be reinstated.

On the strikers' fourth demand, that Melvin have complete control of sales, Crosby's lawyer said Melvin was untrustworthy and should not be rehired. Crosby said Melvin was basically dishonest, he wanted to play profane records, he was sent down in good faith to KPPC and then was subversive with Donahue and made things worse. He was not to be rehired. Avery said he saw no need to rehire Melvin as station manager but thought him a talented salesman and wanted him as a salesman at KPPC.

April

During the second week of the strike, the last week in March, the strikers began planning a benefit to be held on Wednesday, April 3 at Winterland. Being April, a year since the month Donahue started on KMPX, this could be called their first birthday benefit. They advertised it as such and gave it the name Superball. Bill Graham donated Winterland and a half dozen bands volunteered their performances: the Grateful Dead, The Jefferson Airplane, Electric Flag, It's a Beautiful Day, Malachi, and Moby Grape. The Superball was scheduled to run from 6 p.m.-2 a.m. with an admission charge of $5 a seat. Crosby saw an advertisement for it, which appeared in the *Barb* of March 29, showing a couple nude balling and got mad. He thought it typical of the strikers' tactics. It was illegal, pornographic use of the call letters of his station. The Superball was a benefit for them, the strikers, it was not for him. It was not for KMPX.

Shortly after the strike began, Crosby and Hunt had called Miller in Detroit and asked him to come back to KMPX as program director, offering to fly him out. Miller recalled telling them he would come and look the situation over and take the job if he liked it. He arrived in San Francisco on Saturday, March 30 and talked with Crosby and Hunt at the station. As of Wednesday, April 3, the date of the Superball, he had not yet committed himself to take the job of program director or to go on the air. He felt the pressure of public opinion was against it. The strikers knew he was back and considering an offer. McClay told a *Barb* reporter he was afraid that if Miller went back on the air, he might get ads and listeners would figure the strike was over.

Miller went to the Superball at Winterland on April 3, and when the striking staff went up on stage and took bows, went with them. The audience gave him special applause. It looked like he was siding with the strikers. The underground press reported he was.

But the following week, on April 10, Miller took Crosby and Hunt up on their offer and went back on the air on KMPX in the 8-midnight shift, previously Donahue's shift. He opened his show with a comment about it being "the time slot I covet." The *Barb* of April 12 quoted him on that and said that Miller had shown his true colors, a bright shade of chickenshit. Miller remembered a *Barb* reporter interviewed him the night he went back on the air and led him to believe he was being given a chance to tell his side of the story. Then the paper came out and made him look like a fool. On Thursday, April 11, the night the *Barb* of April 12 came off the press, Melvin got a copy and took it into Miller at 50 Green Street. Miller read it and then announced on the air that he was resigning from the station, he would return to Detroit, he had been crucified by the underground press.

The *Barb* of April 19 congratulated itself in headlines for helping the KMPX strikers get Miller off the air. But two weeks later the *Barb* had to report that Miller was back in town and scabbing again on KMPX. He had crossed the picket line in front of 50 Green Street on April 30. His old lady was with him, the *Barb* said, so the strikers thought he must be intending to stay. *Rolling Stone* of May 15 also reported Miller's scabbing and quitting. The *Stone* ran its first article on the KMPX strike in an issue of April 27, along with a picture of members of the KMPX staff assembled in what may have been their initial strike meeting in Prescott's living room. Like the *Barb* and the *Express-Times*, the *Stone* described community response which was largely supportive of the strike, ran figures reporting amounts of money the strikers earned from each of their benefits, and said little about the concerns of Crosby or Hunt.

Hunt, like Miller, had gone to the Superball at Winterland without thinking much about it. Graham had invited him along with Crosby. Hunt said he thought he would go for the hell of it and to see what was going on. He recalled there was joking discussion and argument, and he liked a good argument. Crosby did not attend. He could not see how the invitation was seriously meant for him.

The Superball was the most lucrative of the KMPX strike benefits, raising between seven and nine thousand dollars. Street wrote out the checks giving the proceeds to the striking staff. The chick engineers got $200 each, what they made in a month at KMPX, they said. Donahue and Melvin got $1,000 each. The staff had agreed that the two of them needed funds for flying around the country in search of a way to help settle the strike. Street remembered there was grass and booze provided for the bands at Winterland and the Superball came off well, although Bear, who was supposed to be in charge, disappeared three days before and other people had to do the work. Bear reappeared the night of the show and took a bow with the others. He felt he had done the groundwork by the time he left but it got too much for him near the end, so he took off to the mountains for a few days for a rest.

While Miller and most of the KMPX staff were over at Winterland, Harris walked the strike line alone in front of 50 Green Street; Gleason stopped by with a bottle of wine and told him he thought that was where the story would be. But apparently it was not. Gleason went on over to Winterland and in his column on Friday described the turnout there, making no mention of Harris. He said he had gone and found it a delightfully varied scene. Mingling with the hippies were auto salesmen and executive types, all friends of KMPX. Miller was there with his white hat, one of the heroes of the night. At midnight, the entire KMPX staff filed across the stage, including Tom Dominant, which Gleason said was what the Congress of Wonders called Tom Donahue. It was like a family reunion. Some of the management of KMPX was there. More bands were present than got a chance to play. The bands who did played with their hearts in it. The Dead set off cannons and cherry bombs in one of their pieces.

In the audience at Winterland was one young man who earlier in the day had turned in his draft card in an antiwar demonstration outside the Federal Building in San Francisco. He had grown up in Los Altos near Palo Alto, gone to College in Los Angeles, dropped out, and in the summer of 1967 come back up to San Francisco. He began listening to

KMPX then, thinking the music the station played merged political and cultural consciousness and made it one, as he believed it should be. He thought in terms of media, KMPX was somewhere between what a college was and a newspaper was. At the Superball, he went up and asked Street to announce what had happened at the antiwar demonstration he had attended that afternoon. About 500 people had shown up and a fourth of them had turned in their draft cards. Street did not announce it. The next day he heard on the radio at a friend's house that Martin Luther King had been killed. He thought it must have been on a station other than KMPX that he heard it, because he and his friends felt listening to KMPX during the strike was scabbing.

The day after the Superball, the San Francisco Local of AFTRA, the American Federation of Television and Radio Artists, notified McClay that they deplored the wages and working conditions which had prevailed at KMPX and supported the strikers in their attempt to improve their conditions. McClay had appeared before AFTRA to request their support on April 2. On April 5, the *Barb* came out with the AFTRA letter of support reproduced in full on page three.

The business manager of the San Francisco local of IBEW, the International Brotherhood of Electrical Workers, also a radio employees union, did not urge his local or the San Francisco Central Labor Council to support the strike. He recalled thinking about asking the Council to support it when AFTRA made their statement, but he felt he could not urge endorsement of that kind of demonstration. He had been a union organizer with IBEW for 24 years and he considered the KMPX strike an odd one. It was questionable as a strike. The ordinary man, he thought, did not go on strike without a union. The AFTRA letter of support was to be taken with a grain of salt. The AFTRA people probably felt they should not say no to the KMPX group because members of the group might think of coming to their union later on when they learned they would have to do it by the rules. But then again, they might not. For his own part, he could not see asking the San Francisco Central Labor Council to support a strike with topless pickets. It might get out of hand, and the striking group, he thought, had an overexaggerated opinion of themselves as talent.

From the first week in April through the third, what was done in private meetings, public and community meetings, and negotiation sessions all seemed unsatisfactory to those among the striking staff who wanted to reach agreement on some version of their demands. In mid-April, they still felt justified in their demands but it seemed to Johnson they were losing their ability to imagine the strike could be settled.

By mid-April, a disc jockey by the name of Larry Ickes had been doing the morning shift on KMPX for three weeks. In the eyes of the strikers, he was a full-time scab. Several months before the strike, probably in the end of January, Ickes had sent a demonstration tape to Hunt at KMPX. He was then working at a radio station in Pittsburgh, California. He was 28 and considered himself a professional radio man. In March, shortly after the strike began, Hunt called him in Pittsburgh and offered him a job on KMPX. Ickes said at first he told Hunt no, because the strike was on. In the end of March, Hunt called again. Ickes then came to San Francisco to talk with Hunt and Crosby. He recalled being impressed when he met them that they were not the villains the strikers and the press had made them out to be. They were more like victims. Crosby told him if he took the job he could play whatever he wanted, just so long as he did not say "shit" on the air or play classical album cuts longer than five minutes.

Ickes went on the air on KMPX in the morning shift in the end of March, using as his air name, Larry the Lion. He had to cross the picket line in front of 50 Green Street to go into work. At first the strikers called him over and asked if he realized what he was doing. They said they would not be out for long. Donahue came up to him and said they would really appreciate it if he did not go in and that he was hurting their cause especially because he was a professional. He went in, nonetheless, and came back regularly and crossed the line. The strikers called him scab and other names he felt were abusive. Within two weeks, the line for him became a gauntlet. By about the second week in April, the strikers were threatening him physically and saying things like, "Wait until we give your name to the Hell's Angels. We're not responsible for what may happen to you. We haven't tried anything physical yet." They posted his home phone number with instructions for people to call him 24 hours a day with musical requests, and drove a spike through the door of his Jaguar KE and wrote "scab" on top of it.

By mid-April, Ickes was convinced he was on the right side. He felt Crosby and Hunt were being ganged up on by the strikers and that most people would see that if they talked with Crosby and Hunt personally. He felt if he had not talked with them personally, he would not have come to work at the station, but once he did and then got treated as he did by the strikers, he had to work there. It seemed to him Crosby and Hunt had the press against them in addition to the strikers. He noticed they did not comment to the press and he never understood why.

On Tuesday, April 16, Ickes recalled Donahue came up to him in front of 50 Green Street and told him they were having a meeting on Thursday at the New Committee Theater and he thought the strike would be settled then. The meeting would be open to the public. Graham was going to chair it. For several days before, the strikers had been calling former advertisers, the press, and others concerned about settling the strike, asking them to attend the meeting to show support. Graham recalled he may have initiated the meeting. It seemed to him that by mid-April, the strike was hurting all of them in the music community. It was hurting his dance hall business. Weinberger of Mr. Broadway's remembered he was called by one of the strikers and asked to come to the meeting April 18 to show his support. Whoever called him said the advertisers had a vested interest in getting the station on the air again. Weinberger said of course he would come, advertising on KMPX had totally changed his business.

When Weinberger got to the meeting Thursday night, several hundred people were already there. Many of the former KPMX advertisers had come. Weinberger thought that meeting brought them together and made them realize they were a community. The meeting itself had the feel of a theatrical event more than a strike negotiation session, although there were two sides, the tables were set up to face each other, and Graham chaired it in a serious manner. The New Committee Theater was packed. In the course of the meeting, people cheered and catcalled. Johnson thought Graham was the only one there who was really intent on settling the strike. The owner and manager of Music City said he was told by the person who called him and asked him to come that he could speak his piece at the meeting. He went but he found it boring and left. Van Orden set up equipment to record the meeting for KPFA. Someone held the microphone for the KPFA recording in plain view of the audience, but the proceedings either never made it to the tape or the tape was lost or somehow or other disappeared.

Graham remembered at that meeting the hip types were nice to him for once. Their anticapitalist accusation went Crosby's way instead of his. Donahue spoke on behalf of the strikers. So did Harris and several of the others. Donahue was asked at one point whether he had been fired from KMPX in March or quit. Harris thought that was the first time Donahue admitted in public that he had not formally been fired. Johnson thought even then it was not clear what had happened. It was not clear if Donahue had manipulated them to get them to leave or not. It was not clear if the strike really had to happen.

Then Crosby spoke. Harris recalled Crosby said he would be willing to have the strikers back if they treated him reasonably but they did not. He said he knew sometimes when you were in business, your partners did things you did not like, but they were still your partners. He referred to his early days in radio as a country disc jockey and said he had been a hippie before any of them were hippies, or he was the first hippie, or words to that effect.

When Crosby spoke at the meeting, he was convinced it was stacked against him. The tables were set up so Donahue faced the audience and could speak to them directly while he was sideways to them. The meeting proceeded like a kangaroo court. People were heckling. He expected Hunt would do the talking for him but Hunt was a country boy and got up there and got afraid. He could hardly open his mouth. He was mumbling. They were making fun. So, Crosby said, he spoke. He told them Donahue had hired so many people he could not meet the payroll.

By the time of the meeting of April 18, Crosby felt they were holding a gun behind his head. He had received an eviction notice at 50 Green Street earlier in the day. He felt he had been starving with the station all these years, he was finally about to make it, and these guys came up and said they were taking the station whether he liked it or not. They had kicked the elevator door in the building in. The night he was supposed to go to KPFA in Berkeley to tell his side of the story, they let the air out of his tires. The articles in the papers were all partial to Donahue. Gleason had been on the station for a while before the strike. But when it came, he wrote articles partial to Donahue. Crosby thought he might have talked to Gleason, but Gleason never called him. Caen never called him. The demands the strikers were making were unreasonable. It would have been suicide for him to take them. They were killing him. But he felt he had to stay in it, he had to keep the station.

Graham remembered toward the end of the meeting, he proposed that Crosby and Donahue sit down in private with whoever they needed and man-to-man work it out, they should take the pimple and squeeze it. He offered to put them up at the Highlands Inn in Carmel for however long it would take to settle the thing. He thought they should once and for all find out if the two of them could work together. Crosby refused. Graham thought then there was no hope. He told Crosby he would never advertise on his station again.

Johnson felt that meeting was the last ditch. After it, the staff all realized the strike was going to be a failure. She began thinking it was a result of their taking drugs and being on fantasy trips. So many of them thought the world would be transformed by drugs. After that meeting, she began to feel bad about herself for what the strikers were doing. It seemed to her their tactics were sadistic and it was clear Crosby would not have them back. Sometime the following week, she drove away from the strike line in front of 50 Green Street and never came back. she thought the others held it against her for pulling out. McClay called her, she thought it was shortly before she left, and said another station, KSAN, was sending out feelers to some of the staff. He was thinking of taking a job there, getting himself established, then doing something. She recalled she told him it was every man for himself now and then she hung up.

Melvin said he felt some of the staff's ugliest moments came out in meetings during the strike. He felt the strike was a negative thing while the station had been positive, and you could not keep people going that long on a negative thing. He thought the group of them fell apart during the strike. Street remembered their meetings were demoralizing. As time went on, people individually left. Johnson, she felt, was burned by the strike.

Weinberger recalled after the April 18 meeting, Donahue told him privately he thought they would be back programming soon. Donahue had been making trips to Los Angeles and the East with Melvin to see if he could work something out. At one point, it seemed, he hoped to find money to buy the station from Crosby. Then he hoped to find another station for the group to move to, or a job for himself or a few of them where they might later be able to bring the others on. By mid-April, Donahue had made arrangements to acquire the Beatles film "Magical Mystery Tour" for a benefit showing, but he did not as yet have a theater. By the time of the April 18 meeting, he thought he had buyers for the station. He recalled offering to buy it from Crosby at the meeting. Crosby refused. Crosby was more resistant then Donahue had expected. Donahue thought maybe Hunt pushed Crosby and made him resistant, or maybe it was Crosby's lawyer.

After the April 18 meeting, the question of where the staff might go became urgent for Donahue. Avery suggested ABC. In the end of April, Donahue and Melvin went to Chicago to meet with the vice-president in charge of radio for ABC. They stayed in a motel on Michigan Avenue for four days waiting to see him, feeling duty bound, Donahue said, not to spend more of the strikers' money than they had to. So they ate candy bars and played gin rummy in the motel. They finally met with the vice-president and the man was tempted. He could see the format had been successful in Los Angeles as well as San Francisco, but a deal did not work out. About a month earlier, Donahue had talked with Bill Drake in Los Angeles about possibilities for syndicating the KMPX type of format and had expected Drake would come through with something but that was dead by the time of ABC.

In late April, on his own, prompted by the outcome of the Graham meeting, McClay called a friend of his who was program director at a Metromedia station in Philadelphia. During the first week of the strike, McClay remembered the staff had drawn up a list of all the FM stations in the Bay Area and gone through it considering alternatives. One was the Metromedia FM in San Francisco, a classical music station. So far as McClay knew, nobody talked to anybody about it then. After the April 18 meeting, he called his friend and asked him who was in charge of radio for Metromedia. His friend said it was John Van Buren Sullivan, Jack Sullivan, and he would call and tell Sullivan there was a group from KMPX in San Francisco with a proposal. McClay then spoke to Sullivan on the phone himself. McClay thought he must have been on the phone to Sullivan two or three days a week for two weeks in late April and early May. At first he explained to Sullivan the goal of the strike was for the staff to go back to KMPX. However, should that become impossible, the group would want to go to another station. Then he had to tell Sullivan it did not look like they would get what they wanted.

Much of what the KMPX strikers got in mid and late April were complaints. One listener sent a letter to the *Express-Times* soon after Miller's quitting. Published in the issue of April 18, it ran as a "letter to the world" on the subject of KMPX:

> My dear friends (& especially unto the attention of robert prescott who may please to explain All) re larry miller & the recent Greek tragedy cum Comedy of Errors with emphasis on e. the bear's well known ideal of Love & the Forest and j. the christ's equally well known Forgiveness thing and o every part of joy & friendship which lasts through fault and flaw. HEXAGRAM 37 (THE FAMILY) CHANGING IN THE 3RD 5TH & 6TH PLACES TO 24 (THE TURNING POINT, RETURN). please endeavor to understand this please.
>
> i also add my sad & angry notice: if larry miller did by his action show his "true colors" (a lonesome shade of black and blue) then you too, by your acts have shown your false true-love for Art and Life and Freedom to do One's Thing.
>
> how could larry miller have stayed on/by your side, dusty "super chic" [sic]? you must have hated him all along — even at that great superball reunion (which was it seems for super politics not brotherhood or music or loving cunts and pricks). You must have to so have named him such bad words in the cruel very super printed newspaper.
>
> it is too bad and sad to bear, that we will never have KMPX again.... larry started it so long ago and it seems fitting that he finished it out last thursday night.... it doesn't matter who "wins" the strike — the way & in the light are lost. you see, it was not the programming which really got it — it was and always has been the strong loving, even blessing, intent of

those who played the platters which made ex-your ex-station a way to live in & with.

but now you people hate too much...; think i shan't again believe your recorded words of love ... we all lose ... you are bitter exiles, the scabs are inept creeps, larry (out of the mud springs the lotus) miller is going back to Detroit ("by day i make the cars and by night i make the bars") city, and the listeners (i am one, hello) turn off their radios and learn to live alone.

god bless us all and may we yet find all the love we need.
Catessa

May

On the evening of Saturday, May 4, Crosby sent a telegram to Avery informing him that a meeting of the board of directors of Crosby-Avery Broadcasting would be held on Monday, May 6 at 2 p.m. in the office of Crosby's lawyer in San Francisco. On Monday, May 6, in an outcome of that meeting, Crosby sent a telegram to Avery informing him that he had been removed as president and general manager of Crosby-Avery Broadcasting.

Months later, when Crosby finally asked his lawyer to resign from the Crosby-Avery board, he thought maybe he should have taken him off earlier instead of taking Avery off. But at the time they were in the midst of a strike. The strike was technically against Avery's interests as well as his own and Avery was acting as if it was not. There had been one meeting of the board where the next day Donahue did things indicating he had known what went on. Crosby said he presumed then that Avery was the culprit, but it might have been his lawyer. Crosby liked Avery.

Stefan Ponek had feelings for neither of them. In May 1968, Ponek was a disc jockey on KSAN, previously KSFR, the Metromedia FM in San Francisco. He knew the KMPX strike was on. In April, he felt he had started being opportunist about it when he convinced the KSAN general manager to let him do more of what the strikers had been doing on KMPX by extending the length of a rock music program he did on Saturday nights. He had started the program in January in a style like that of KMPX, calling it the Underground Sunshine show. It ran for an hour each week. In April, with the KMPX strikers out, Ponek felt there was advertiser support for him to extend the program to four hours.

Doug Dunlop, a salesman with KSAN, sold the time on Ponek's show. When the staff of KMPX went on strike in March, one of their initial publicity handouts had been a list of the names, addresses, and phone numbers of advertisers who had gone off the air on KMPX in support of their strike. They urged people to patronize these advertisers. Dunlop said he got hold of a copy of the list and started calling on the advertisers asking them to place buys on Ponek's show. By late April, he had done so well selling Ponek's show that its expansion was possible. The station manager, Reid Leath, was for it. Other than on rare occasions, he said, the station had not before received requests for time by advertisers.

Dunlop wore three-piece suits when he went out selling time on Ponek's program in April and May. The merchants he had gotten from the KMPX list were mostly hip types. He knew he looked straight but he felt the fact he was young helped him. He was 22. Also he was enthusiastic. When he sold time to the Middle Earth clothing store on Stanyan Street, the manager said he reminded her of her son. Several of the former KMPX advertisers called Harris and asked if it would be all right with him for them to advertize on Ponek's show. Harris recalled telling them okay, that KSAN was the best thing going in town.

Ponek remembered in April inviting the KMPX strikers to appear as guests on his show. He extended an invitation to all of them but Prescott and Bear were the only ones who came. Probably in the end of April, he suggested to Leath that they hire some of the striking staff from KMPX. Leath said he was worried about trouble with Metromedia because the station was not making money and he did not want to do it. Then one Monday, probably Monday, May 6, Ponek said Leath called him into his office and asked him what he would do if he were in his position. What would he do if he had to hire a group of people he could not control? Ponek said he told Leath he would take them but buy them off one at a time. Leath said to go ahead and do that, go ahead and hire them and he could be program director. He also told him it had to be a secret.

The day he talked with Ponek, Leath had just come from New York where he had gone for a managers' meeting with Sullivan. He had previously talked with Sullivan about expanding Ponek's show and about possible combinations of rock and classical programming on the station. But after this meeting, he knew they were going to change the format of his station and they were going to change it entirely. It was only a matter of setting the date and making the local arrangements.

Leath did not know at the time that earlier in the first week of May, Donahue had been to New York and talked with Sullivan. McClay had gone East on vacation and had set up an appointment for himself and Donahue, following on his phone conversations with Sullivan. Donahue flew East, joined McClay, and met with Sullivan at 975 Park Avenue. Donahue recalled their meeting was brief. They got along well and concluded it all in one day. He was impressed that Sullivan had really listened to the music.

The proposal resulting from their meeting was that Metromedia would take some of the staff from KMPX onto KSAN and change the KSAN format from classical and lively arts to rock. Donahue would come on as program director. The present KSAN air staff would all be dismissed except for Ponek, who would stay on as a disc jockey. Donahue did not like the prospect of Ponek's staying on, as that would leave the KMPX staff short one man, and he did not like it that KSAN was a combo operation with the disc jockeys doing their own engineering. That meant there would be no positions for the chick engineers.

When he returned to San Francisco, Donahue talked with Melvin. Sullivan had not made a formal offer, but it was likely. They talked about how the strike had been getting hard and how they would have a problem splitting shifts if they went to KSAN.

McClay, not yet back, was also disappointed by Sullivan's reluctance to take the whole staff. Sullivan seemed to want to pick off the cream. But he had an enthusiasm for KMPX. Sullivan felt KMPX was exciting, like WNEW had been in New York in the late 1940s when he first went to work there. Sullivan had become manager of WNEW and subsequently an executive of Metromedia and was already a grand old man of radio when Donahue and McClay met with him. He was one of radio's spiritual leaders, one who businessmen in the business called a philosopher, a program man, the kind who did not care about details. He recalled having been to San Francisco and Los Angeles in 1967 and having heard the music the bands were playing and the two stations programming it at the time. It seemed to him the first step forward since the bands of the 1930s and 1940s, it had honest lyrics, and he could see it hit the 20-29 age group, the growth group for the next decade.

To Sullivan, it made eminent sense to take the staff from KMPX when Donahue and McClay came to him in May. He felt you learned in this business there were damn few really original personalities as opposed to imitators and realness was important. Knowledge of music and records did not mean anything without it. What had made WNEW was that people listening could really identify with the air people. That was also what made KMPX. The people in San Francisco were into the new music and it was right for their lifestyle. It did not have to be Sullivan's lifestyle for him to see it had appeal. When he met with Donahue, he liked him. He was high on Donahue. He liked his philosophy, he thought Donahue was the gospel father of progressive rock radio. He felt good about the whole thing. Then he had to tell the president of Metromedia he wanted to change the format on KSAN by hiring the group from KMPX. The president of Metromedia had reservations. He said what in hell did Sullivan think he was doing?

Sullivan said all he could tell him was this was for San Francisco and for that market it was right. He had been hammering away at the 20-29 age demographic for all their stations, so had the corporation's research department. He felt the president of Metromedia had reservations not because it was bad marketing to change the format, but because it was new and different.

The president of Metromedia was an investor, not an operator in radio stations, so of course he would not understand. He had originally been in food brokerage, representing grocery manufacturers to supermarket operators, had bought into Metromedia in 1958 when it was the Metropolitan Broadcasting Corporation, and had become its president in 1959. In the nine years since, Metromedia had developed a reputation as a fast growth company which bought red ink and made money off it. In May 1968, the company owned a dozen radio stations, half of them AM, half FM, and four television stations, all in major markets. Their other principal holdings included Foster & Kleiser, the nation's largest outdoor advertising company, a transit advertising company, a direct mail advertising company, a documentary film company, The Ice Capades, and *Playbill*, a theater magazine. In January 1968, *Forbes* magainze quoted the president of Metromedia as saying the whole idea was to reach the customer where they could.

In May, Metromedia already had one radio station with a format like that of KMPX, WNEW-FM in New York which had gone full-time in its format in February, having started part-time in October. KMPX had started part-time six months earlier, in April, and gone full-time in August. The general manager of WNEW-FM, George Duncan, said Sullivan initially encouraged him to put in the rock format by suggesting he find out about the revolution in music. Duncan had been a salesman under Sullivan at WNEW-AM.

After Donahue and McClay met with Sullivan in May, they went to see Duncan. Duncan recalled Sullivan asked him to talk with them. When he did he found out that Donahue had done a format change on KMPX in San Francisco like his on WNEW-FM, but with a different approach and unbeknownst to him at the time. Duncan said he felt their approaches were different in that Donahue had introduced his change on KMPX in opposition to abuses of top-40 radio, he had been idealistic about it, while on WNEW-FM, they were more technical and analytic in their thinking. They thought about changing the format in terms of what would make good radio and good business sense. Their thinking was that album cuts had an intellectual appeal and that the programming should be compatible with the style of the music. Any jingles they played would have to be compatible. The style would have to work as a whole if it would work at all. Duncan felt this was different from how it happened at KMPX. At KMPX, Donahue also eliminated jingles but he did it with defiance.

Duncan recalled the KSAN manager Leath came to New York for a meeting shortly after Donahue came. He advised Leath at the time to expand the number of hours they were playing rock music on KSAN and suggested he call Donahue about it in San Francisco and then call Sullivan in New York. He did not let on that he knew Sullivan would not refuse if Leath called and said he wanted to hire Donahue.

In San Francisco, after Leath gave him permission, Ponek started trying to buy off members of the KMPX group individually. One of the first he approached was Pigg. Pigg remembered meeting with Ponek and telling him he could not make a decision without consulting the group. Bear remembered Ponek contacted him that second week in May and asked him to keep their meeting a secret. Ponek seemed nervous about it. Bear said he wanted badly to go back to work, to go back on the air after all this time, but he told Ponek he could not come without the group. Melvin recalled meeting with Ponek in a Foster's restaurant near KSAN. Their meeting was supposed to be secret. Ponek brought the KSAN sales manager with him and the two of them wanted to know what kind of business the KMPX group would bring. Melvin said he told them all the accounts would follow. That knocked them out. Ponek suggested a group of them come without Donahue, that he would be program director. Melvin said he told them they would all have to come together. In the end, they agreed to set up a meeting with Leath to see what could be arranged.

Ponek made offers to Pigg, Bear, Melvin, McClay, and Prescott. It seemed to him each of them was afraid of being a bad guy in the eyes of the others. Ponek felt he was putting his own neck out to do it. He had to keep his advances secret from the classical staff at KSAN. So far as he knew, he and Leath were the only ones at the station expecting the format change. When it came, the classical staff would be canned and they would hate him for his complicity. He told Leath he had been turned down by each of the KMPX strikers he approached, that they each said they had to check with the group. Leath said he felt it was blackmail.

Probably on Thursday, May 9, the KMPX staff still on strike held a meeting. Pigg recalled he was the first to say he had been approached by Ponek for Metromedia. Then the others spoke up. Bear thought each of them said they had told Ponek yes, but they could not come without the group, and that at that meeting they agreed it was all of them or none.

Bear remembered that meeting especially because he spoke out against Donahue. He felt the strike morale was ready to fall apart and that it was time for the staff to go back to KMPX. He told them he had been talking with Crosby and Crosby was willing to take some of them back. They had gone out initially because of Donahue but now they had been out too long. He said for himself he was ready to go back, he did not need a Big Daddy.

Donahue recalled at that meeting Bear announced he had been talking to Crosby and that Crosby wanted to take some people back. Donahue thought it was at the same meeting they worked out a package to go to KSAN. They had to decide on a group because they could not all go and they would be short of a man because of Ponek's staying. They agreed that Bear would not be in the package. It seemed he did not want to come.

Bear thought it happened differently. It was not in the meeting of Thursday, May 9 that they excluded him but over the weekend. On Saturday, from what he was told, Donahue met with a small group of people closer to him to put together the package for KSAN. Donahue told this group that Metromedia did not want Bear to come, or that Bear did not want to come, and they could not take the chick engineers. Harris called Bear after the weekend meeting and told him he was not being included in the package but he thought Bear should be going with them if he wanted to. Bear said he told Harris he wanted to go. Harris then said he would speak to Leath at KSAN about it. Before he went, Harris met with McClay. McClay was against it. He acted scared, Harris said, he thought it would hurt the group. But Harris went anyway and told Leath he thought Bear should be part of the package from KMPX. Leath said he would bring it up to Donahue.

Leath subsequently met with Donahue. Donahue recalled he bullied Leath into letting him take one more man. They agreed on a package and afterward he called Bear. Bear remembered Donahue called and asked him if he wanted the all-night shift on KSAN. On KMPX, he had the midday shift. Donahue told him they had been planning to give the all-night shift to the ladies, or to one of the ladies, but Leath had objected so now it was his if he wanted it. Bear said yes, he would take it. He did not tell Donahue or make public at the time that Harris had called him the weekend before and met with Leath on his behalf. He did not make it public because he wanted to protect the station. He did not want to hurt the group.

Donahue recalled Leath objected to the ladies taking the all-night shift in a second meeting he had with him. In the first, there were other things to settle. One was that Ponek expected to be program director. Also he was disturbed they were not going to let Melvin be sales manager, but there seemed nothing he could do about it. Then there was the problem of losing the girls.

It seemd to Leath that for a small station, KMPX had one of the biggest staffs he ever heard of. They came and asked him to take their heavyweights and offered to give him the rest for free. But he knew people did not work for nothing. He told them he could take only a selection. Then in all the commotion, word got out and people assumed there would be a format change on his station before he formally announced it.

Ponek thought Donahue probably went to New York after Leath went there, that he must have gone after he found out they were making offers to buy off his staff individually. He must have arranged in New York to get himself put in as program director and then came back and talked with Leath to work out details. That must have been why Leath told Ponek he would keep him on as a disc jockey but that Donahue would be program director. It was only after that Ponek first met Donahue.

Ponek attended the last meeting of the KMPX strike on Monday, May 13 where Donahue announced that the staff could move to KSAN and that it was arranged for the following Tuesday, May 21, Donahue's fortieth birthday. The group assembled moved and voted to end their strike and divided up what remained of the strike fund. Johnson came back for the meeting. She remembered Donahue told them KSAN was set up for combo operation so they would not have positions for the chick engineers, but that he would talk them into setting it up for engineers and then rehire them. She did not believe him. She told him at the meeting, "I don't believe you," then said she was just going to leave and do her own thing. She felt the pressure of the drug scene was too much, and she felt she had been used by Donahue. Street said she felt Donahue had led them to believe the whole family would go on to greater heights. Then it seemed he said fuck the whole family, eight of us are going to back to work. It was arranged for five of the disc jockeys and three of the salesmen to start at KSAN full-time on May 21. Some of the others would come on weekends. There was a suggestion that the chick engineers and a few of the rest form a company of their own to produce commercials.

Prescott recalled feeling good about the prospect of going to KSAN because he got a job. He would be morning man on KSAN as he had been on KMPX. Nothetheless, he felt the move to KSAN was second best. The staff had wanted to go back to KMPX. Part of the dream was to be independent, to work for themselves and not be interfered with. They had wanted to get rid of what interference they had from Crosby at KMPX. In retrospect that seemed not much. A big corporation would be worse.

Melvin thought they were caught at a disadvantage by the Metromedia offer. They wanted so much to go back to work. But it was tucking their tails they went. They were selling out going over there and they knew it. The justification for it was it would be better than the streets. Just Thursday of the week before, he had gone on the air on KMPX as a guest, reviewing the strikers' grievances once again and arguing against the side of the scabs who were on the air. He told them they would not escape the problems the first staff had. Melvin had been offered a management job at *Rolling Stone* during the strike and had turned it down. He would be a salesman at KSAN.

The KMPX strike was over. The underground press reported it toward the end of the week of May 13. The *Barb* said when word got out, Miller announced on the air on KMPX: "Well now they have their own station, we have our own station, and you have your radio." The *Barb* also reported on a showing of the Beatles' film "Magical Mystery Tour," May 12 and 13, in the Straight Theater on Haight Street. The showing was a benefit for the striking staff of KMPX, arranged some time before. It was their last strike benefit. The turnout was impressive but the film reviewer for the *Barb* was not sympathetic. The poor could march on Washington, he said, but the heart of this community was with people who complained they were not getting enough bread for playing records. The KMPX strikers had wanted creative freedom, and what that amounted to, he said, was freedom to play cut three, side B instead of cut five, side A.

Crosby felt relieved that the strike was over. He would hang on although people were urging him to sell the station. The strikers would be happy somewhere else. It was ridiculous, he thought. Donahue had tried to make him look like a wealthy man. Donahue had told people he was a rotten capitalist so-and-so. He had not said much in reply, he did not want to tell everyone his problems. Now Donahue was going to Metromedia, a giant corporation. Crosby believed in capitalism within reason. He did not believe in giant corporate capitalism or huge inheritances. He thought if the strikers were really looking for independence, they had it at KMPX. Going to a huge corporation there would be no freedom at all. It was money they were after.

PART II:
LEGITIMACY

Chapter 5

KSAN
(May-August 1968)

The staff now must learn to operate in the context of a large corporation. There are new rules with which they are expected to comply and these are made clear in introductory meetings, new appointments, and corporate attempts at control.

Sullivan thought there was lemonade. Hamilton said there were lemon and cherry pies, she had made them. There was also coffee. There may have been lemonade. Toward the end of the week of May 13, Sullivan came out to San Francisco to speak with the staff from KMPX who were going to KSAN. He met with them at Donahue and Hamilton's house on Scott Street. Donahue arranged it so the staff came in shifts. Sullivan recalled he told the staff that Metromedia was the kind of company that let their stations do their own thing. It let them specialize. He told them he believed their kind of station was good radio. He said he was convinced of it. Philosophical discussion followed. That, he felt, was what the meeting was for. He told them he had been president of the Radio Division of Metromedia for only a year and a half and he spoke of how he felt about his role.

On the morning of Tuesday, May 21, about half a dozen of the former staff of KMPX found their way to the offices and studios of KSAN on the fourth floor of 211 Sutter Street, a ten-story office building on the corner of Sutter and Kearny Streets in downtown San Francisco. An insurance company had offices next door to the radio station on the fourth floor. One stepped out of the elevator, crossed an entryway, opened a door opposite, and walked into a large lobby of a room in which there were desks used by the station's salesmen. Off to the right of the lobby was a small production studio. Down a narrow hall to the rear was the main broadcast booth — a combination control room, announcing booth, and record library. Turning left, in the rear of the station, was a large room that served as the office of the program director, and behind it, separated by glass, a studio used for commercial production. Past the program director's office, in the left rear corner, was the office of the station's general manager. Coming back down a hallway on the right was a small kitchen, then the office of the traffic director, then the shop of the chief engineer.

Prescott arrived at 211 Sutter Street shortly before 6 a.m. on Tuesday, May 21, in time to do the morning shift. When he opened the door to the radio station on the fourth floor, one of the first things he saw was the wallpaper, Metromedia wallpaper, a muted multicolored print, undistinguished, yet somehow awful. When the rest of the staff arrived later, they found themselves faced with the same paper and moving in on the station's former classical staff. Eight of the old full-time staff were leaving. Six were staying: Ponek, Dunlop, the promotion director, the receptionist-sales secretary, the chief engineer, and the bookkeeper, traffic director, and office manager, Donna Campbell. Campbell, of all of them, had been there the longest. She had been hired in 1966 by the man who owned the station before Metromedia bought it.

Ponek said it was strange when the KMPX group first came over. They were celebrating, here, in this previously dead station. The receptionist from the old staff said the group from KMPX seemed to be happy about coming. They were dressed as hippies. She thought they did a better job advertising the format change than Metromedia. Campbell felt upset when they came. The old classical staff was still around. The classical people had been told by Leath they were fired only a few days before the format change went into effect. They were promised two weeks severance pay and asked to leave at once. Some of them went on the air with impassioned bitching the weekend before the change and they were said to have taken some of the station's equipment with them when they left.

The promotion director who stayed on said she did not hear about the format change from Leath until a few days before it happened. Leath may have made some nonspecific remark about it earlier in the month, but she did not actually know until the day Donahue walked in. Then she got calls from newspapers asking about it. Still she had not been told. She thought that was why the classical staff who had to leave were bitter, not because they had to leave, but because of how it was done.

Leath was uneasy about the change. He had felt from the first that he might not be able to control the group from KMPX if he had to hire them. From what Donahue knew, his fears were not unwarranted. In their initial meetings. Donahue said he sized up Leath and decided he was crazy. It was not Donahue's fate to work for seemingly sane and competent station managers. He sometimes blamed it on the nature of the business. This one he would have to placate as he had the others. He would give Leath a daily problem, to keep him happy, to keep him out of their hair. Ponek said it was doubtful Leath would have gotten in their hair very much anyway. His preference was to withdraw. He would pull the blinds and keep to himself in his office in the back of the station.

Donahue opened his first show on KSAN at 6 p.m., Tuesday, May 21 with two cuts from an album by Blue Cheer, three by Aretha Franklin, and two by Dave von Ronk. There were commercials in-between. Hamilton was in the studio with him. She kept a list of the music he played. The staff was going to have to keep music logs at KSAN starting May 31. At KMPX, those who wanted to keep a record of what they played did and those who did not did not.

On Thursday, May 23 at about 11:30 p.m., Donahue played "The Pusher" by Steppenwolf on his show. "The Pusher" had at one point been forbidden airplay on KMPX.

Donahue recalled one night early in June, after the staff had been at KSAN about two weeks, he was on the air. Hamilton was with him. She was reading meters. It was about 8 p.m., Stone was there with his girlfriend. Pigg was there with his girlfriend. Johnson and Street were in the production room. Three men came in. One was the FCC lawyer for Metromedia from Washington, the second the director of personnel for Metromedia from New York, the third the general manager of KNEW, the Metromedia AM in Oakland. It seemed to Donanue the three of them were upset. They wanted to know who Hamilton was. He told them her engineering license was on the wall. They wanted to know why there were girls in the production room who were not employed by the station. They implied this was no way to run a radio station. The next morning when Leath came to work, he was mad. Donahue then called Sullivan in New York. Sullivan said he had told those dumb bastards not to go out there and do a thing like that.

Tom Dougherty, the FCC lawyer for Metromedia, remembered going out to the West Coast early in June to check the situation at KSAN. The format change was causing trouble as he had expected. Complaints were coming into the FCC and to Metromedia in New York. A columnist in the *Chronicle* was objecting. Two days after the format change went into effect, the FCC had initiated an inquiry into the outside business activities of Donahue and his staff. They could not forget he was a payola jock. Dougherty said he went out to the Coast to look into the backgrounds of each of the people from KMPX. He took it as personal detective work. The night he and the director of personnel for Metromedia and the general manager of KNEW stopped in at KSAN, they were checking on operations of the station. It seemed to Dougherty Leath did not have the situation under control. He could not control it. Donahue was running the station. But Donahue did not have the backing of the ownership. Dougherty said he talked to Leath first thing in the morning. Then he called Sullivan in New York and said someone should get the situation under control.

Leath kept his position as KSAN general manager to the end of June. In the start of July, it was announced that he was taking a three-months leave of absence. He was replaced by the general manager of KNEW, Varner Paulsen, who called KSAN "the FM." Paulsen was medium sized and bald. He had been with Metromedia long enough to feel committed to the company and with KNEW as general manager for just over two years. Sullivan said he thought Paulsen knew the San Francisco market and could wear two hats and oversee KSAN on an interim basis without much trouble. He thought Donahue had the station in hand. Sonner or later someone would come and ask to manage it independently. It might be Donahue. It might be someone else. When you had a good organization, you did not worry about such things.

Pigg thought Paulsen might best be described as World War II hip. When Paulsen took over, he called a meeting of the KSAN staff to introduce himself. The meeting was held in the offices of KNEW in Jack London Square in Oakland. Prescott recalled Paulsen started that meeting by telling the staff he was an exploiter and was going to exploit them and what they had in the name of Metromedia. Boucher recalled Paulsen tried to lay down the law at that first meeting, and he was impressed with how Paulsen was different from Crosby. You knew exactly where he was, he was against you. Donahue thought Paulsen came in like a Nazi. He took the staff on a tour of KNEW as if to show them what a real radio station looked like, as if in a couple of months they could turn KSAN into something like it.

Paulsen said he called the initial meeting at KNEW so the staff could get to know him quickly. He wanted them to know he was serious and a professional broadcaster and that he wanted KSAN to be a professional station. He believed the progressive rock format they had was not a fad, but it had yet to be justified to society. He thought the group from KMPX had previously been coming to work to have fun. They had no discipline, no responsibility. He had to be concerned for Metromedia about the property and the bottom line and he wanted them to know that. In the initial meeting, he felt a wall between himself and the staff. He was a broadcaster. His purpose was to enlist clients to advertise on the radio station. He was interested in the money. He told the staff if that made him a prostitute, he was. He told them they would have to generate enough business to make Metromedia's investment pay.

Alan Stone, one of the disc jockeys who came part-time from KMPX, remembered Paulsen did not get much backtalk from the staff after his speech. They felt now it was going to start. From here on they would just be working. Their perfect radio station was not going to be. When they had started at KMPX, Stone thought, they had no history. Then suddenly there was a past, even at KMPX, and people began getting stuck in it. Now at KSAN they had Paulsen, a small-timer who thought he was a big-timer, a businessman.

Ponek thought Paulsen was militaristic but that beneath it he was limp and the staff knew it. He could give orders but he knew he would run into trouble if they were not funneled through Donahue. Stone recalled in the beginning at KSAN, Donahue asked the staff not make him tell them what to do. Paulsen thought Donahue was Mr. Nice Fella in the eyes of the staff. He thought they liked Donahue. They listened to him. Paulsen said he respected Donahue.

Donahue said he and Paulsen had an implicit agreement about running things at the station. Paulsen would be mostly at KNEW, and he would do most of the running of KSAN. Paulsen would come in and talk with Campbell and Donahue would ask him questions to make him feel in a position of authority. There was no conflict, Donahue said.

A batch of congratulatory telegrams arrived for Donahue and the staff when they went on the air on May 21. The following week, on May 31, the *Barb* ran an advertisement showing the Congress of Wonders pointing to a radio. "Old-timey family radio" had returned to the air, the ad said, with "the originals" and some additions, and listed their names. The *Express-Times* of May 23 ran an article reporting that Donahue had been emcee at a rock concert in the Santa Clara County Fairgrounds the Saturday before the staff began at KSAN. There were long-haired people at the concert, the article said, but most of the audience was 15-year-old white kids, about 8,000 of them, lots of short sleeves, some bermuda shorts: "These kids came from Scouts, Sunday Schools, mowing the lawn for chores and maybe getting a pony for Christmas. And they're going straight out of that towards the world of Pigpen and Janis."

On Monday, July 1 at the Avalon Ballroom, the staff of KSAN held their first cocktail party. The purpose, according to the station's promotion director, was to publicize the format change. The party was scheduled from 6 p.m.-8:30 p.m. and given the name, Free-Form Family Freakout. Creedence Clearwater Revival and West played. Little Joe's Electro Luminescence did the light show. Busy Line Caterers provided food. The food cost the station $2,200. The music and light show were nonexpense items. The ballroom was obtained on a trade for $175 of air time. Advertising agency, public relations people, retail clients, newspaper columnists, and record distributors were invited. About 1,000 people came. Champagne and hors d'oeuvres were served.

On July 23, Donahue spoke to a conference of the National Association of FM Broadcasters at the Fairmont Hotel in San Francisco. His subject was adult rock programming.

The KSAN promotion director thought in the beginning when the staff from KMPX came on, they expected to run the whole station. They wanted to be in charge of themselves like they had been at KMPX. They were not about to be run by Metromedia. Leath could not handle that. Paulsen did better. He tried to be hip, and he was, she felt, more hip than she was. He was only there part-time, but he had more control than Leath did being there full-time. After he came on in July, he wanted to avoid controversy. Donahue also did. They wanted not to offend people. They wanted to keep the station out of trouble. There were three groups who might have given them trouble if provoked, but aside from the initial protest of the format change, the first, the straight people, did not say anything. The second, the hip people, did not have to. The third, the FCC, did not know what was going on.

The staff from KMPX seemed to be settling in at 211 Sutter Street by the end of July. The station's office space had been somewhat rearranged and there were plans for remodeling. By the end of July, the rock staff had been on the air on KSAN two weeks longer than they had been on strike from KMPX. Many of the advertisers from KMPX were back on with them. They seemed to have listeners. They had more visitors hanging around the station than Paulsen felt should be there. In a memo of July 22, he warned: "Again — No Visitors — or incense. Business visits only. On weekends and evening hours the place is off limits to everyone except staff and immediate assistant. Please don't make me prove I mean it."

By the end of July, there were plans for Donahue to move into the left rear office previously occupied by Leath. The room in the center in the rear would become the station's record library. Campbell would move to an office downstairs on the third floor and act only as bookkeeper and office manager. The receptionist and sales secretary would take over traffic. Booths would be installed in the front lobby area for the salesmen.

The sales manager would have an office downstairs near Campbell. A shower had been torn out of a small room off to the right of the front lobby and that space became a supply closet. Because Paulsen had an office at KNEW in Oakland and was coming across to KSAN a couple of hours a day at best, he would share an office with Campbell on the third floor. Also on the third floor, Metro Radio Sales and Metro TV Sales would have new offices. The Metro Radio Sales office sold time on nonlocal Metromedia radio stations and some independent stations.

Paulsen referred to the plans for remodeling in a monthly report he sent to Sullivan in the beginning of August. The report covered the month of July and was organized according to Metromedia policy for monthly reports from all of their radio stations. It had a cover summary by the general manager and sections by the department heads. The monthly reports were said to be for purposes of information and record within the corporation. Duncan said he thought people said things in their reports they might not want to have said again and he felt the reports could easily be misinterpreted by someone who did not know the context. Harris said he thought the reports were justifications. Their purpose was to give the company confidence. They were a station's lease on life for the next month.

In the engineering section of the KSAN monthly report for July, the chief engineer said he had installed automatic logging equipment during the week of July 15 to free the air staff from the task of taking transmitter readings so they might concentrate more fully on their air product. But after a week's experimental use, trouble had developed in the automatic clock and the unit had to be returned to the manufacturer for repair. Early in July, facilities were installed for the master recording of tapes to be used on automation equipment at other Metromedia stations. On July 22, an audio line was installed betwen KNEW and KSAN to enable direct sending of news reports for KSAN use. Additional cartridge equipent for playback purposes was ordered. The station began 24-hour operation on July 17 and eliminated the six-hour down period on Monday mornings. By use of two separate transmitters, they could now schedule maintenance on either transmitter during regular hours.

In the programming section of the July report, Donahue noted that the ABC news and the Sunday night symphonies had finally been eliminated from the station's programming. The rock music format was now on full-time. The staff had played "Music from Big Pink" by The Band for three and a half weeks prior to its release in July. In the first four days of its release, an initial shipment of 2,000 albums was sold out in San Francisco and the record company was back-ordered 20,000, and KSAN had been the only station in town playing it. The station's record librarian, Donahue said, was cataloguing their collection on a large Rolidex and was half way through. All in all, he felt, the staff was settling in and the station would prove itself in upcoming ratings.

The business manager's report for July said the station was operating at a profit loss for the month that was about half as much as it had been for the same month the year before. The sales gross was $13,631, almost four times what it was the previous July.

A new sales manager, Jerry Klein, started with the station in August, replacing a sales manager who had been hired by Leath before the format change. Klein left a job as a salesman for a television station in Sacramento to come. He had straight hair and was 24 and his father was sales manager of KNEW. They brought in this kid, Melvin said, the son of somebody. From then on it was downhill. The salesmen from KMPX realized they had sold out. They sold shit. They did not care. It was everybody split and out for his own ass. Everybody scatter and run and make it while they could. Toward the end of August, Melvin quit his job as a salesman with the station. He said it was because of this kid Klein they brought in to be sales manager. Laughlin looked at Klein, saw he had short hair, wore button down shirts, said "gee whiz" when he should have said "far out," and never forgave him.

Dunlop said it was his opinion that Melvin should have had the sales manager job. But Metromedia was skeptical of putting any of the freaks in management positions. Paulsen said he put Klein in as sales manager because of the bad history of the station with regard to seriousness and being professional and because, being Metromedia, they had to be concerned about bottom line. Klein was young. Paulsen had respect for his father. He thought his father might be able to help him in the agencies, and he felt Klein did not have the phoniness, the contemporary culture, the others had.

When Klein came to work at KSAN on August 5, he thought it would be exciting, a challenge. He was going to manage a group of people for the first time. Then he walked into the situation and it shocked him. He did not know what was going on. Laughlin was a gun-carrying revolutionary and telling everyone off. Paulsen gave him the assignment of going through each salesman's list of accounts and taking their advertising agencies away from them, then following them up himself. He divided the list of accounts into five instead of four and took one. Aside from himself, the four salesman at the station at the time were Melvin, Laughlin, Harris, and Dunlop. The third of the Western triumverate from KMPX, Towle, had gone to be sales manager at WBCN, a progressive rock station in Boston, which had offered their sales manager job first to Melvin and Melvin had passed it on to Towle. Towle said he thought the few of them who had sold time on KMPX had become instant experts. They were the only people who had done it before and done it successfully.

Chapter 6

KSAN
(September 1968-March 1969)

The disillusionment of the staff is especially noticeable during these months. It accompanies efforts to broaden the station's pattern of support by appealing to listeners and advertisers unfamiliar with the station's style. There is one dramatic firing over an incident concerning commercials and this is at once both a loss and an opportunity for a statement about the nature of losses the station now incurs.

McClay said he felt divorced from the station during the period Paulsen was general manager. Beginning in November, McClay was switched to the all-night shift. He had been doing late afternoons. He was still running Tempo on the side. His shift on KSAN was, he felt, a job. Nothing seemed significant to him about the station except his show, and it was difficult to keep awake even for that. Paulsen had been appointed for an interim term but that was little comfort. No one knew how long his interim would be or what would come next.

On Saturday, September 7, Melvin married Mimi Farina, the sister of Joan Baez, at the Big Sur Folk Festival at Esalen Institute in Big Sur. Hamilton turned 21 in September. Melvin stopped working at the station as a salesman and went to work for Voco at Mercury Records. Voco was then a staff producer with Mercury. In late October, the KSAN record librarian, Lynne Hughes, who had been sales secretary and then record librarian at KMPX, resigned. Donahue said in his monthly report it was to pursue her singing career. With Voco's help, she got a contract with Mercury. Also in October, Street went on the road with Voco as a recording engineer. She felt she realized then that radio was a business. They had been a bunch of idealists at KMPX. Then their dreams got shattered real fucking fast. That was when she went to Voco.

Laughlin started doing a regular weekend talk show on KNEW in October. Harris thought Laughlin's getting the talk show went back to the first week in July when Laughlin told Sullivan off. Sullivan had stopped by the station for a visit. It was in the afternoon. Harris said he walked in and found Laughlin telling Sullivan it was assholes like him that were the enemy, people who drove Buicks, lived in the suburbs, and were executives in big corporations. Sullivan sat there with his white hair. He was the man who had done it, who had brought them in off the streets, and there Laughlin was telling him off. Harris said he stood by the door of the office they were in and listened, but he could not stand it and left. He thought they kept arguing all afternoon and that Sullivan must have realized Laughlin could talk, because he ended up giving him the talk show. Sullivan did not recall the incident.

Laughlin said it was in Paulsen's office and that two weeks later Sullivan got him a job with Donahue doing a talk show on KNEW. He had just come back from a weekend in the mountains with Melvin when he walked into the station and met Sullivan. He was high on dope. He and Sullivan had an argument about radio. It was not so personal as it was a clash of life-styles and an argument about what this kind of radio should be. The key line was when Laughlin said the staff at the station and their audience were not the kind of people who did so and such and drove a Buick, and Sullivan said he drove a Buick.

Laughlin guest hosted his first talk show in KNEW in the end of August under the name, Travus T. Hipp. The *Express-Times* ran a news item on August 28 saying family radio was taking over the talk show of Van Amberg this week, from 7 p.m.-9 p.m. on KNEW, with Travus T. Hipp presiding, and that Van Amberg's show normally had an audience of one to two thousand middle-aged matrons and off-duty butchers. Anyone who wanted to rap and lay his story on these people was invited to call in. On October 12, Laughlin again went over to KNEW to host a talk show, this time his own, on a regular weekend basis. The *Barb* of October 18 said that Laughlin used the seven-second delay repeatedly to cut out obsenities coming his way on his show but he did not catch all of them, and KNEW received an unusual number of complaints from listeners. Copies of some of the letters of complaint had been sent to the FCC, but the KNEW program director said he was not worried.

Paulsen, in his October report, said he felt all was well with KSAN and that Donahue was a damn good man, totally committed and mindful of his future potential with Metromedia, but he wished Donahue would lose some weight. With respect to national sales, he said he had talked with the head of Metro Radio Sales and recommended they hire an FM specialist. Ponek, in the October report, said the staff had been attempting satire in the station's news with the use of free-form collages. He thought they were in good taste and he hoped the station would hire a person to do the news full-time in 1969. Donahue said that the air staff was refining their music programming and attempting to maintain a balance and they had received a flood of new releases during October. About a third of them made it into the record library after review by the staff. Thursday nights between 8 p.m. and 10 p.m. they were doing two hours of Dylan plus a guest artist. On Wednesdays between 10 p.m. and midnight, Pigg had a program of oldies on.

Billboard ran a feature on progressive rock radio in its issue of October 19. The article said the format was stirring the radio world across the United States, and that KSAN in San Francisco had moved out of nowhere to become the number of one FM station in its market in the June-July Pulse, and this was just two months after dropping classical music for progressive rock. The progressive rock format, *Billboard* said, was the radio success story of the year, and not only in major markets. There were at least 63 stations across the country playing it. The two biggest successes were KSAN in San Francisco and WNEW-FM in New York, both Metromedia stations.

The station got a new chief engineer in October, replacing the one who had stayed on from the classical format who, Donahue felt, hated the guts of the rock music staff. Richard Nixon was elected President of the United States in November. The campus of San Francisco State College was closed down indefinitely. There were executive changes in Metromedia, and Sullivan became vice-president in charge of corporate relations, a newly created office for the purpose of coordinating properties for a proposed merger of Metromedia with Transamerica. The headquarters of Transamerica were located in San Francisco, the Metromedia headquarters were in New York. The staff at KSAN started making comments on the air about the proposed merger soon after it was announced, especially concerning a giant pyramid Transamerica was planning to build in the City.

Sullivan's place as president of the Radio Division went empty in November. Paulsen addressed the station's monthly report to no one in particular. In it noted that there had been no provision made in the 1969 budget for a separate general manager for KSAN and he thought this should be considered now because it would eventually be necessary and he had enough on his hands with KNEW. Two other FM stations in the San Francisco area were on the air in November with formats like that of KSAN and there was a possibility of a fourth. The salesmen felt the others in competition for local retail accounts.

In December, at least half the staff came down with the flu. Toward the end of the month, Donahue attended the Bill Gavin radio programming convention in Las Vegas. Duncan of WNEW-FM also went and remembered asking Donahue then why he did not take over and run the damn thing, run KSAN. Donahue told him he knew he could do it but he was not ready. He did not think he would like it. He was into show business. He was a ham.

Fifty queen-sized bus posters showing the Congress of Wonders pointing at a radio were up on San Francisco buses in November and December. There was a wave of pre-Christmas spending and commercial time on the station was sold out for about two weeks. Donahue went around with Klein making presentations to major advertising agencies using a tape they had put together to introduce the station. Klein recalled they would talk and play the tape, then he would show numbers on KSAN. He would point out that the station was the dominant FM in the 18-34 age group and compare it with the top-40 stations KFRC and KYA by saying it was one-fifth as expensive. He would pitch the agencies that if they really wanted to reach the youth, they should buy KSAN too.

Klein said he thought of himself as a middleman. He stood between the hip world and the establishment. The station's traffic director said she thought Klein was not so much in the middle as he was incidental to what was going on. He was wishy-washy compared with Melvin and Laughlin. They were super salesman. Klein said he thought he was involved in a slow educational process with respect to the advertising agencies. He would start with the buyers and try to move back through the account people to the client. He felt he was handicapped because people in the agencies were afraid their clients would get mad if they found out KSAN was a revolutionary station, and many of the clients had not bought FM at all before. And he got no help from on top in Metromedia. The head of Metro Radio Sales was against the format and there was little incentive for the national reps to sell it. He was supposed to goose Metro Radio Sales into selling it. But he knew if the reps were pitching KNEW for the San Francisco market, they were pitching a 50-plus age group and they were not likely to knock themselves out running around saying, "Hey you want to buy some teeny boppers too?"

When Klein and the sales staff did get national accounts, the air staff objected. The station was not a radio station at the time, Klein said. It was a voice of the revolution. The air staff had a "they can pay for us" attitude toward Metromedia. They had a confused image of Metromedia.

A new salesman joined the staff in January, to replace Laughlin, who was planning to leave in February. Paulsen wrote in his monthly report, with Laughlin going, all the salesmen would be straight, except for Harris who was only slightly hip. Aside from some of the talent, Paulsen said, they had by the end of January phased out quite a few of the avant-garde characters. But the station still needed a full-time manager. Donahue did a good job minding the store but he was a poor administrator. His interests ran hot and cold and he had lousy follow-up.

The ratings were holding. The union contract was up as of the end of January. The director of personnel for Metromedia would meet with the Local beginning in February: "I don't see any problem here," Paulsen said, "except, perhaps, that IBEW may not want to represent the group."

Donahue, in January, spent more time than he liked working on sales. He made visits to the agencies along with Klein but he never liked working with salesmen. Maybe it was their greed for money. They seemed to have to have it. In January he also had to deal with the new president of the Radio Division, a businessman who had succeeded Sullivan, and with the head of Metro Radio Sales who was against the format. He tried to explain to the head of Metro Radio Sales once when he came out for a visit that KSAN was what KSFO used to be, a middle-of-the-road station but in the middle of a different road, a freeway. The new president of the Radio Division was said later to have heard his remark and said it was ridiculous.

Donahue felt it was a continual struggle after they lost Sullivan. Once in a while, he would call Duncan at WNEW-FM. He always wanted to fight all the battles but Duncan knew you did not win wars by fighting battles and would say so. Duncan was an ex-Marine. Paulsen was an ex-Marine. It seemed to Donahue possibly a requirement: to succeed in Metromedia you had to be an ex-Marine. Donahue said he felt there was no prospect for his promotion in Metromedia. Duncan had suggested he try for the manager's job but he did not think he stood a chance to get it. The staff at the time, the staff from KMPX, was not as Christian toward these people as they might have been. Donahue said his theory was they would be better off proseletyzing than fighting the opposition, people like Klein, the head of Metro Radio Sales, and the new president of the Radio Division. But it got to him after a while. He stopped making tapes to send down to KMET in Los Angeles during the winter. He thought about leaving.

In February, Klein went after the Hastings Clothing account at Gross, Pera and Rockey, the biggest media agency in the City. He finally got them to buy the station. They sent over spots. But the spots they sent, Klein said even he could tell, should have been aired on KSFO. They had a father on them berating his son for wearing long hair. Donahue caught it first and told him, "Come here and listen. Send these back." Klein then went to the agency and offered to have the station produce a commercial for them but they refused. The station ended up losing the account.

Aside from criticism of the commercials, it seemed to Klein there were three main topics of conversation at the station during the fall and winter. One was politics, how bad the establishement was. A second was drugs. For the first six months he was there, August through January, Klein thought the staff were in favor of all drugs. Then they started being down on heavy drugs, especially Pigg. A third topic was nature and ecology. That got going in the fall. Also there was nudity. Klein thought Bear did some of his shows nude. Klein knew he had differences with the staff from KMPX, but he felt after his first six months he was accepted by them, and considering where he started from, that was something. Beginning in December, he had to hire new salesmen for the station. He thought he made a mistake with the first one he hired, but he wanted to hire someone straight. He had to turn down a lot of guys who said they came from the Haight-Ashbury and had dropped out for a while but wanted to get back into the business world.

On the night of Friday, February 7, Bear played two commercials simultaneously on his show. The following week, on Tuesday, Paulsen fired him and issued a memo to the staff:

To: Entire Staff
From: Varner Paulsen
Date: February 11, 1969

I will start this off with "Peace," but what follows is at the opposite pole.
    We are spinning our wheels in so many areas that I am becoming increasingly concerned about what we are as a station, and where we are going.
    We have had several instances of air men knocking commercials or otherwise not performing as believable salesmen. That's our business, "gentlemen," to sell goods and services that have been approved by the Sales Department and me. I check every order. Beyond that point, you do the selling. If you have something against a business school, or a product buying time on the station, see the salesman, Jerry or me — but don't use my microphone to shaft the account. I have a quick remedy for non-professional broadcasters active in commercial radio who don't know where they're at.

A week and a half later, on Friday, February 21, Bear distributed a 10-page letter to each member of the KSAN staff and stapled with it a copy of Paulsen's memo of February 11, suggesting that both his firing and Paulsen's memo were evidence of what their once free-wheeling radio station had come to.
    Bear said he had come into the studio on February 7 shortly before 10 p.m. to do his show. The air shifts had been rescheduled in November so Bear now followed Donahue. He came in with a recording of the

Shastakovitch Symphony No. 5 in his hand. Donahue was finishing his show with a cut which may have been sung by Lynn Hughes. At 10 p.m., maybe a minute after, Bear spoke into the microphone: "Good evening. This is Edward Bear. First off, let's clear out the lightweights." Then he played from the Shastakovitch Symphony. The music, he felt, was intense. Donahue left the studio.

Later during his show, Bear played a commercial for Dana Morgan and a commercial for Sounds Unlimited at the same time. It was an accident, he said. He pressed a button for one commercial cartridge to play and when it did not play, he pressed another button for another cartridge, at which time, maybe a second later, they both came on together. Also during his show that night, he announced that certain records were selling for $1.44 when the price listed in the commercial copy book in front of him was $1.00.

Harris heard Bear on the air mess up the commercials. Bear suspected that Harris and Donahue manipulated Paulsen into firing him. He also thought Paulsen was predisposed because he felt the station was out of control at the time. The moving of McClay to the all-night shift in November, Bear thought, had been Donahue's way of saving McClay from becoming their first sacrificial victim, the first one of the KMPX staff to be fired from KSAN to allay fears of Paulsen, and higher-ups in Metromedia for all he knew, that the situation at KSAN was out of order.

Bear said he felt in retrospect he must have insulted Donahue personally when he started his show by saying he was going to clear out the lightweights. He thought Donahue must have felt clearing out the lightweights referred to him and the kind of music he played. He talked with Donahue afterward and Donahue said something to the effect that he was really a big triangle teetering on one point. He had to have an extra sense about him, a third eye, to watch out for things that might make him fall.

Harris had felt it was unfair the previous May when Bear was not included in the group from KMPX going to KSAN. But now in February he felt Bear got what was coming to him. Harris heard Bear say, "let's clear out the lightweights." He heard him mess up the commercials. He thought Bear did an angry show, and that Bear was angry because he had not been able to get a switch shift he had wanted to go see a concert and it had, for other reasons, been a bad week for him. Bear said yes, he was having difficulties. But he tried to keep up his show. He had gotten a massage earlier in the evening that Friday to get himself in shape. When he arrived at work and came on with Shastakovitch and his "clear out the lightweights" line, he was trying to get himself into a mood to do his show. He wanted it to be a serious show. He wanted those listeners who preferred not to go on a heavy trip to be forewarned he was into one. His purpose was not to deprecate Donahue.

He opened with the Shastakovitch Symphony No. 5 because it was the most intense piece he could think of. He followed it with Pink Floyd mixed with some electronic music. Maybe he was less than charming that night, he said in his letter to the staff, but he thought that was part of the essence of free-form radio. Even if it was not, he thought he had done enough good shows to be allowed a bad one occasionally. He thought he had been careless in playing the two commercials at once. But he felt it was in its own way aesthetic. It might be considered an attempt at multilevel subliminal delivery. He knew the station could make good on the lost commercial time.

Harris felt Bear did not like the station's commercials in general. If he was not putting them down for bad taste, he was correcting grammar or spelling on copy the salesmen wrote. One or two of the spots Bear messed up were spots Harris had sold. He left Bear a note about it, and on Monday night, Bear left him a reply:

> Dear Whitney:
> Glad to hear that my listenership is so finely tuned to what I'm doing, be it sponsors, salesmen, management or listeners — anyone with a stake in it — that there can be useful comments.
> Seems like someone caught an error or two of mine. Fine. I am happy to hear Benny is selling all mono at a buck and I'll be happy to rap it down. When I was there I saw signs posted above record racks saying $1.44 and such and that was what I said. Is giving someone a make-good spot as serious as two blue demerits or is it simply to bed without dessert?
> Sorry about the Dana Morgan and Sounds Unlimited Duo. It was actually unintentional but I decided to let it go once I realized I had hit both due to the content of the one of them. For an accident, I thought it went off rather well.

Bear was given a three-hour notice when he was fired by Paulsen on Tuesday night, February 11. Wednesday night, he called the office of the Berkeley *Barb* and told a reporter about it. The *Barb* of Friday came out with an article headlined, "KSAN Sacks Bear," run on page five beneath a picture taken the previous spring showing a group of the strikers of KMPX in front of 50 Green Street. According to the article, Bear had phoned the *Barb* office Wednesday night to say he had been canned Tuesday by the KSAN station manager Paulsen. The official reason was he ran two commercials simultaneously the previous Friday. However, he told the *Barb*, it was really just him. He had been running a show which presented music and craziness. He had tried to do free-form radio and speak his mind, but apparently that was dangerous.

KSAN had beautiful ratings which appealed to advertisers, Bear said. But now they were going to dump the people who built the ratings, the hairy revolutionaries. They were going to clean up the station for the advertisers.

In his letter to the staff, Bear said although he had been told he was fired because he aired two commercials simultaneously and misquoted a record sale price, the truth was different:

> I was dumped by an absurd series of misinterpretations, confusions, and accidents that combined with super-sensitive egos, paranoia piques and a general communications breakdown. It really was all so strange. I would much rather have been fired for reasons that would inspire me to battle the forces of cruelty and injustice. As it is, it's more like a sad, dumb joke — rather a monument to pettiness than to evil.

Further in his letter, Bear said he noticed no one had mentioned on the air in the preceding two weeks that he, Edward Bear, one of their number, had been fired. That could haved been because no one cared or had nothing to say, but he felt it was because they were afraid:

> It couldn't have been that nobody cared that I was bounced and had nothing to say. What was sad was that no one felt free enough to. And I don't mean crusades or eulogies or even a defense of me — but merely the outfront reporting of a relevant fact. It was relevant that I was fired and it wouldn't have breached any code of faith to say it. If saying a lot would have been superfluous, then saying nothing, or what amounted to nothing, seemed lame and untrue to the spirit of it all.

He went on to remind the staff of KMPX and the strike. He said he thought nobody won that strike, and that those of them who had delivered free-form radio into the consciousness of America, who could only have done that with the force that came from genuine freedom, were being absorbed into the lukewarm goo of commercial compromise. They were being told, "play your music, freak, but keep your thoughts to yourself — be professional." At KMPX, he said, it had not been necessary for them to lie or tell half-truths in order to sell accounts. Doing business at KMPX required dealing with people who were content to buy your air time but not your soul. It required dealing with people who did not need your soul.

The *Express-Times* of March 11 carried an article on Bear's firing and quoted from his letter to the staff and from the memo issued by Paulsen. The *Express-Times* also quoted Donahue as saying that working for the big companies was like deciding to go to bed with someone. You did not turn around and say "don't touch me" after you were in.

Donahue voiced similar sentiments in an article he wrote for *Record World* in January which appeared in the *Record World* issue of February 17. Donahue began it by saying in the last two years he had done a lot of writing and talking about the so-called underground or, as he preferred to call it, free-from radio, and he knew the issues it often raised:

> I guess the most oft-repeated question that inevitably arises in its plastic coated paranoia is "How long will it last?" The question is based on the assumption that "they" (whoever the hell they are) will not tolerate the continued existence of anything so beautiful and so free. Before the conversation is over you're sure to hear that large corporations will fear its image, that crass commercialism will scar its beauty, that governmental restrictions will destroy its freedom.

His own opinion was:

> There will always be a segment of your audience that considers any kind of advertising that goes beyond head shops and boutiques as a form of sellout. If this is the guide line then we have indeed sold out, as has everyone but The Fool on the Hill (when he stays on the hill). I do not believe that the commercial per se is evil.

In the final analysis, he said:

> The continued success of this kind of radio is dependent upon a very simple element — good taste in musical selection. The disc jockey in Free-Form radio today has better understanding of today's music than anyone else in the business. Not only is he playing it on his show four hours a day or more, but he is in constant contact with the artist and the consumer while others in the business are involved primarily with their own product.
>
> I like disc jockeys that are essentially groupies, who love their music and take it home with them and are involved with it to a degree that approaches fanaticism. I don't expect them to like the same things I do or play the same records or construct their shows in the same fashion, but it is only with this intensity of interest and understanding that you can hope to pursue the thematic ideas that are so attractive to an audience and generate the kind of musical enthusiasm that has been gone from radio since the early days of Top 40.

A good disc jockey should be truly an artist, creating a kind of pop art that has an immediacy that is untouchable yet at the same time disappears as fast as it is created. He signs his name to an echo and a memory is the only residue of his achievement. Like all artists he creates for himself. If he is a great artist his recognition will be in those memories.

On January 31, the 530 Washington Street Corporation terminated by default Leon Crosby's lease on studios and offices for KMPX at 50 Green Street. Crosby relocated the station not far away at 495 Beach Street. On February 25, Crosby's lawyer in San Francisco sent to his FCC lawyer in Washington a copy of an agreement whereby the National Science Network Inc. agreed to purchase both KMPX and KPPC from him. The date of the agreement was February 15, the total purchase price, $1,300,000. Most of it went to cover debts Crosby had incurred during his ownership of the stations. Crosby's accountant said Crosby made a net of about $200,000 on the sale and that half of it went to his former wife. A quarter of it went into escrow and much of the rest went to cover personal debts.

Hunt said it got to the point at KMPX where he knew they could not survive when they were caught between the income tax guy and the phone guy, stalling them off with each other. Ickes said things got so bad they could not afford to pay a janitor to take out the trash at 50 Green Street. Old lunches and things accumulated. Crosby was concerned and for some reason assumed Ickes would clean it up. Crosby would stop by every so often and say to him, "Larry the Lion, would you mind giving the place a lick and a promise." That was what he would say, "Larry the Lion, would you mind giving the place a lick and a promise."

Ickes thought he remembered the day Crosby realized things were so bad he would have to sell the station. They had already moved to Beach Street. Crosby walked up and said, "Larry the Lion, how would you like to come with me and get an ice cream cone?" Then they walked over to Swenson's.

Hunt remembered negotiating the agreement with the National Science Network to sell the stations. National Science drew up the final papers. The day in February when Crosby was supposed to sign the papers, Hunt said Crosby told him he was not going to do it. He did not want to give it up. Hunt said he told Crosby he hated giving it up too. He had been fighting for over a year. But there seemed no other way.

Crosby said the day he had to sign the papers, he almost could not do it. He did not want to sell. He had held onto it for so long. He had starved with it. He had stuck out a strike. Hunt and his lawyer, a different

lawyer since the strike, had to take him over. They had to hold him up so he could sign the papers.
Hunt thought it was too bad. In the end Crosby had to sell the station at a loss. Crosby gave him a check for $1,200 on the sale and said he was grateful to him. Hunt had slept with him nights on the floor of KMPX during the strike. Hunt stayed on at KMPX for about six months after National Science took over running it. But he felt they did not want him. When he left, he went back to Michigan. He stayed out of radio for a while after that, for four or five months. He just wanted no part of radio.
Ickes stayed on at KMPX. He kept his morning shift, and his name, Larry the Lion. It seemed to him after Crosby signed the purchase agreement with National Science, a lame duck period set in at the station. It lasted until National Science got approval for ownership the following November. Crosby was supposed to be the manager but National Science was signing the checks. A representative of National Science came in every few weeks and looked around. The staff was paranoid. They did not know when things would change or where it would come from next. They expected almost every day to find some notice saying they were all fired.

On March 27, Paulsen issued a memo to the KSAN staff on the subject of his replacement as their general manager:

To: Entire Staff
From: Varner Paulsen
Date: March 27, 1969
Subject: New Man In Town

As some of you may have suspected, we have been looking for a full-time General Manager for quite some time. Starting Monday, April 7, your new Manager will be Willis Duff. He's an excellent choice; I recommended him in December.
Willis served as Program Director at KLAC in Los Angeles when it was a music station in 1963. He continued in that capacity when the station went "Talk" in 1966. He's a great guy, a good broadcaster — and "with it."
Willis left us in 1968 to become Manager of WHDH in his home town, Boston. He has resigned that post for personal reasons and rejoins Metromedia Radio at KSAN.
You'll be interested to know that Willis also programmed Progressive Rock in Boston and has a feel for today's music and other social enlightenments.
I hope you will give him maximum cooperation.

I enjoyed working with you all, but the time has come to concentrate complete attention to the challenge at KNEW. I'll be around to see you from time to time.

Peace,
Varner Paulsen

Chapter 7

A VOICE OF THE REVOLUTION
(April-August 1969)

The corporation now meets the style of the station with a manager more appropriate to it and conflicts between "us" and "them" become more intimate within ranks among the staff. At the same time other stations with similar formats are beginning to draw followings and the station's competitive situation is changed. Yet the fact of the others' emergence indicates an expanding market for the format and contributes to the station's legitimacy. The staff continues the effort to become less exclusive by appealing to those unfamiliar with the station's style. This in turn exacts concessions from that style. The results are indicated by changes in attitude among the staff.

Willis Duff had a conscience, Donahue said, but he needed to be reminded he had one. He would get lost between the two sides, Metromedia and the staff. Then Donahue said he would tell him, "Willis, you're the secret hero of the revolution." Hamilton thought Duff did not know there was no revolution. In that way maybe it helped to be dumb. The KSAN traffic director, previously the receptionist and sales secretary with KSFR, remembered Duff called the staff to a meeting when he first came. He was straight in how he dressed but casual and he put himself down for being so square. He talked about what he had done in Boston and said he would be open here. He told the staff he wanted to learn and invited them to come speak to him individually to tell him what they wanted to do and what they expected of him.

When the staff walked in for their first meeting with Duff, the traffic director thought Duff must have noticed they were cold, super cold. They were expecting another Paulsen. But Duff, as he talked, did not seem like another Paulsen and after the meeting people were hopeful. They went to see him individually, at least all the disc jockeys did. They did, Duff said, and almost every one of them asked him for a raise.

Duff was 32 when he moved from Boston to San Francisco to become the third general manager of KSAN. He had a wife and two sons. He had been born in Texas, gone through the Air Force Academy, attended three universities in Texas, and written a 60-page handbook on talk radio. He had hung around radio stations as a kid and had held his first radio job at a station in Bonham, Texas where, he said, he did everything from sweeping floors to producing and announcing. The radio jobs he had after that were mainly at Southern stations. There was one for a short time in Providence, Rhode Island and then KLAC in Los Angeles. WHDH in Boston was the first time he had a general managership. KSAN was the second. Duff wore glasses, had a moustache and a trim beard and short brown hair when he came to KSAN.

    He arrived in San Francisco in mid-April, a week later than Paulsen initially expected but in time for Donahue's wedding. On Sunday, April 13, Donahue and Hamilton were married. Their reception was held at the house of The Jefferson Airplane on Fulton Street near Golden Gate Park, a large Victorian on the edge of the Haight-Ashbury. Many of the KSAN staff attended. Prescott remembered the weekend of Donahue's wedding was the weekend Duff's wife arrived. She went with Duff to the wedding and at the reception Prescott saw both of them eat the food. He felt sorry for them because no one had told them the food was dosed. Later Prescott heard that Duff and his wife went up to Mount Sutro or some place like that and the lights of San Francisco were unlike any they had ever seen. Duff said he and his wife did get stoned at Donahue's wedding. They went to a point opposite the fort under Golden Gate Bridge after it and the lights of the City were beautiful.

    Donahue recalled the food was dosed at his wedding. But as far as he knew people got high in a good way. They were not expecting it and therefore had not built up fears which might make it a bad experience. The champagne punch was also dosed. But, Donahue said, he did not do it. He did not believe in dosing people without their consent. A group which hung around with The Jefferson Airplane was said to have done it. The May 31 issue of the *Stone* carried on page three a six by eight-inch photograph of Donahue and Hamilton during their marriage ceremony. He wore a suit and she a wedding dress. She was holding flowers. In view behind them was one of the salesmen and one newsman from the KSAN staff. A caption beneath the picture read, "All of Tom Donahue's wedding." There was no accompanying article.

    Donahue left KSAN at the end of the week after he and Hamilton were married. Duff said he implored Donahue to stay. Donahue said before he left he tried to make the transition easy for Duff. He had considered leaving the station a couple of months before Duff came but had not. He felt a responsibility. This kind of radio, this station, was his baby. He was willing to turn it over to Duff. But he wanted the kid to be raised right even if he was not going to be around to take care of it. He knew

Duff expected to turn the station into a well-running business operation. He spent what he felt was a lot of time with Duff going over things. Duff said Donahue was his educator and mentor. Prescott recalled there was maybe a week overlap in April when Duff and Donahue were both at the station together. After that, from what he knew, Duff went out to Donahue's house for advice.

Prescott said he felt sorry for Duff when Duff started. About a week after Donahue's wedding, Prescott went to a promotion party at the Fillmore given by a record company to promote one of their artists. Duff came in and it seemed to Prescott he made an ass of himself. He just walked over and introduced himself to people saying he was Willis Duff of KSAN. Prescott said he went and talked with Duff after Duff's first meeting with the staff at the station. Duff was looking for someone to be program director. Prescott had been operations manager toward the end at KMPX but he did not want the position at KSAN. He thought, while talking to Duff, that Duff was intelligent. He had an Eastern academic background and he was analytical and cool in how he was approaching the station. But he did not seem to have the feel for it. It was hard to tell if he would be good or bad for it.

Harris remembered on Friday of Duff's first week, Duff came to him and asked where he could get some hip clothes. "Take me to one of your advertisers," he said. Harris, Pigg, and Klein then took Duff to Changing Faces on Upper Grant Avenue and got him some clothes. They had dinner at the Minerva and later that night went to a concert at Winterland. Bear was there and Duff met him. Harris said he hyped Duff on Bear then.

Duff remembered letting them show him around when he came. He let them lead him. He sort of liked it.

Duff said Paulsen told him he had a high opinion of the staff at KSAN but that the corporation might bug him about superficialities. That was his first warning that the corporation might try to get him down. Paulsen said he believed by the time he left and Duff came on that the staff had more discipline than they had when he started as general manager 10 months before, and the discipline had to do with what they considered were their roles. Duff said he noticed immediately that the staff had an us-them attitude toward Metromedia. They seemed to expect autonomy within the corporation. They were fetishist about drugs. Also they were megalomaniac. They thought they were great. They expected to be number one in gross ratings. At first he disagreed.

In the end of April, when Donahue left, Pigg got Donahue's air shift and Ponek got his program director position. There seemed little question among the staff that Pigg was the one best suited to take the air shift. But the appointment of Ponek as program director was thought by many of them to be a mistake.

Ponek said he suspected Duff picked him to be program director because he reminded Duff of a younger version of himself. When Duff came into the job, the station had virtually no sales manager and no general manager. Donahue was leaving, and Donahue, Ponek thought, was the type of leader the country allowed before the industrial revolution castrated all the men. Yet Donahue had a habit of side-stepping trouble and letting the axe fall somewhere else. It seemed to Ponek, Donahue pulled out of the radio station just at the time they were going to have to go through growing pains. The pressures were there. They were going to have to go revolutionary or commercial, one way or the other.

They were handicapped by the loss of Donahue, and Donahue was a hard act to follow. Ponek felt Duff as general manager and himself as program director could not help looking second-rate in comparison with Donahue, and Duff, in comparison, was too afraid of what he was dealing with. He had this air of constantly juggling eggs. As a result he could never really assert himself. He would torture himself with indecision.

Duff did not remember feeling so tortured, or thinking of Ponek as a younger version of himself. He did remember thinking Ponek was the only one on the staff who wanted the job as program director after Donahue, and there was nothing at the time which indicated Ponek could not do it.

Ponek still felt like an outsider at the station when Duff came on. When he took the job of program director, he thought it might be a means to prove himself to the staff. Ponek was 30. He had been born in Vermont, gone into the Air Force after high school and worked on armed forces radio. He had moved out to Santa Barbara in 1965, attended the city college, learned about drugs, encounter groups and mysticism, and worked on KMUZ-FM, an automated slush music station. His wife got pregnant in 1966 and shortly after that he started looking for work in San Francisco. In August 1967, he had taken the job with KSFR and Leath had put him on a split shift, 6 a.m.-9 a.m. and 12 p.m.-2 p.m., and told him to play classical music and on the morning shift to be funny.

After naming Ponek as program director, Duff made several other new appointments of staff. In July, he made Campbell business manager. She had been the station's bookkeeper and office manager. Campbell felt her appointment was not due to Duff, however, but to her own initiative. She said she got the position of business manager by blackmailing Duff. She told him two weeks after he came that she wanted the position and would have it or leave. Her appointment came through formally in July. Before her appointment, the station did not have a business manager of its own. The business manager of KNEW served also as business manager of KSAN.

As business manager, and previously as bookkeeper and officer manager, Campbell, unlike most of the rest of the staff, felt and openly said she felt that her primary loyalty was to Metromedia. Before coming to work at the station, she had worked for a cement company in San Francisco, and before that for a real estate company in Vancouver. In each place she had felt her loyalty was to the company she worked for. She had gotten the job with KSAN in 1966 when it was KSFR, before it was bought by Metromedia. At the time she was looking for work as a computer programmer. She had just completed a programming course. She was hired at KSFR to be the station's bookkeeper, librarian, and traffic director. She could do all three jobs because there was not much volume then. Campbell was 29 in the summer of 1969. She was, she felt, compulsive about her work and self-critical. She had always been that way. Her appointment as business manager brought her a slight raise and the status of a better position. But it did not mean any basic change in the kind of accounting work she did.

For Campbell, working for KSFR and then KSAN was working for a business. She knew people in radio liked to think of their jobs as more glamorous. A secretary or a traffic director might plan to get a license and one day go on the air. But for her it was different. When she interviewed applicants for clerical positions with the station, she tried to make it clear to them they would be taking jobs in a business.

When the format change first went into effect, the chief engineer with the classical staff had asked Campbell why she was not against it too. She was not, she said, because she did not care about the music. At home she listened to softer music. For all she knew, the format of the station might change again. The rock group might leave. But she would stay. Her loyalty was to the corporation.

As Campbell understood it, Paulsen was supposed to have been general manager for only two months until Metromedia could find a regular manager. But his term dragged on. Donahue pitched for the manager's job when Paulsen left but the odds were against him. He was an unknown commodity to Metromedia and they were wary of him.

Klein also thought Donahue pitched for the job of general manager when Paulsen left. Klein said he pitched for it himself. Then Duff came. Duff seemed to Klein to be bright and interested in everyone and he let them know it. But Klein left KSAN in the end of May. He felt good at the time he left about what he had done at the station. He felt in terms of sales, they had things under control by April. He left because he had antsy pants. He was getting bored with the work and a couple of television rep firms had pitched him for 50% more than he was making at KSAN. Duff asked him who he would recommend to follow him as sales manager. They went over each of the salesmen. He told Duff he thought Harris was to indelicate for it.

But Harris was the one Duff finally appointed to replace him in June. Duff recalled he thought Harris had more experience than any of the other salesmen at the station. Harris said he was not sure whether to take the sales manager's job when Duff offered it to him. As a salesman, he felt he had freedom. A salesman was only responsible for his own production. As manager, he would have to take on the whole of it. But to say no to the manager's job would be to say no to his future, to the corporate possibilities, and he wanted to see if he could do it.

Before appointing Harris as sales manager, Duff asked Dunlop if he wanted the job. Dunlop said he told Duff no. He felt at a disadvantage in terms of age. Maybe he would want the sales manager's job in the future but not now. Now he was having a good time selling everything he could get his hands on. He was telling prospective buyers he felt they had something new at KSAN, something special. They were not proven yet. They did not have much in the way of numbers. A buyer would have to like the station to go with it. He would ask them to have faith in the station, trust it for a while. He would tell them they were helping it along.

The third salesman Duff considered for the sales manager's job was Rick Gardner. Gardner had been hired by Klein in February and had taken over Laughlin's accounts, most of which were record companies. Laughlin, Gardner said, was the music specialist. Klein recalled he hired Gardner because of Gardner's enthusiasm. Gardner acted like he wanted to sell and he was logical. He could make convincing presentations. Klein thought people would buy Gardner. It was no secret, when people bought a radio station they were really buying its salesmen.

Gardner was 29. He had grown up in San Francisco and gone to law school for a while. A friend of his who knew Klein had told him Klein was looking for a salesman at KSAN. He was selling Fuller Brush at the time and trying to put together a rock'n'roll booking agency. He had almost gotten the agency together. They had some groups and ideas about assembling a soul show. When he started selling time on KSAN in

February, Gardner felt what he was doing was largely education. The market was there. His job was to let people know about it. But with the small retail businesses, he sometimes got in the position of doing more than selling the station. He got to know their business as a whole. In the case of one hairstyling shop, he got to the point where he was giving advice on their advertising generally, and once on an employee's salary. He felt a personal satisfaction when it happened like that.

In the spring and summer months, Gardner was in the habit of following up some of his calls and visits to clients with letters. He kept copies of his letters in a looseleaf notebook where he also kept copies of contracts he signed. One letter he wrote on June 24, for the purpose of getting the Ponce Beauty College as an account, presented a station history. It began by describing how Donanue and other renegades from top-40 radio had formed KMPX. Their approach was exceptional, Gardner said, because they considered the listener an intelligent, sensitive being. Now at KSAN, research was showing the listenership to be of above average intelligence and income and interested in themselves and their careers. KSAN was of extraordinary importance to this listenership. "We are," Gardner wrote, "a community or milieu station offering credibility. People look to us to tell them what to do, what to believe, and what to buy. KSAN is the kind of station to which people stay tuned, often for weeks, without change. We are not background, we are listened to."

On June 19, Gardner wrote to Woodstock Ventures Inc. in New York concerning an upcoming Music and Arts Fair urging Woodstock Ventures to consider KSAN in their national promotion. Gardner thought he had been the only one at the station who had gotten excited when Charlie Tuna came in early in June talking about Woodstock. He thought maybe it was naiveté that made him believe Woodstock would happen. He subsequently kept in contact by phone with Tuna and arranged a deal for the Diners/Fugazy Travel Agency to sponsor a KSAN charter flight from San Francisco to Woodstock. The travel agency had an office across from the station on Sutter Street. They were underlings, Gardner said, and they liked the station. He drew up a contract whereby time bought by Diners/Fugazy would be matched equally by KSAN for the purpose of promoting a KSAN excursion to the upcoming Woodstock Festival. Scheduling of all spots would be left to the direction of the station pursuant to effecting the strongest promotion possible.

Beginning the fourth week in July, Woodstock excursion announcements were aired live and ad lib on the station with words to the effect:

Here's a wonderful way to spend your summer vacation. Fly to the Woodstock Festival near Woodstock, New York, home of Bob Dylan and the Band, for the greatest collection of groups ever together in one place. Fly American Airlines with Stefan Ponek and a mad, wild group of KSAN people on a special KSAN Woodstock Festival Flight which is being arranged by the Diners/Fugazy Travel Agency (note: Fugazy is pronounced foogáy-zee).

The Festival runs for three days: Friday, Saturday and Sunday, August 15, 16, and 17.

Some of the artists who will be appearing are: Friday: Joan Baez, Arlo Guthrie, Ravi Shankar, Sly and the Family Stone; Saturday: Canned Heat, Creedence Clearwater, the Grateful Dead, Janis Joplin; Sunday: The Band, Jeff Beck, Blood, Sweat and Tears.

Thirty tickets were sold to listeners by means of the Woodstock excursion promotion. Six tickets were given to the KSAN staff for use on the flight, paid for by American Airlines in trade for commercial time. Two of the six were given away by the station to listeners who submitted their names on postcards as entries in a contest. There were about 3,000 entries. The name of the winner was drawn on the air by John Fogerty of Creedence Clearwater Revival.

Bear was rehired by Ponek for weekends in May. Ponek said in his monthly report he rehired Bear after much deliberation over the causes and reasons for his firing in February. Wes Nisker replaced Ponek as news director and the job was made full-time. Nisker continued to produce satirical news collages and they began calling him "Scoop" Nisker, a name Nisker said Pigg game him. In mid-May, a park in Berkeley became the subject of local controversy. Nisker covered it for KSAN. Ponek felt Nisker got his reputation then and that Nisker was like Abbie Hoffman but more loveable. There was a sense of humor on the streets and Ponek felt Nisker caught it. In his programming report for May, Ponek mentioned Nisker's coverage of People's Park and said he thought because of Nisker, KSAN outdid Berkeley's own KPFA, which was known for its coverage of local demonstrations.

Duff said when he made Nisker full-time news director in May, he felt Nisker could articulate a radical conscience without seeming cliched. Nisker was 27. He had almost gotten a Master's degree in journalism at the University of Minnesota. He had visited Israel and stayed for a while on a commune, and had done some street theater. The first time he produced radio news collages was when he started working as a volunteer at KSAN in the fall of 1968.

The psychedelic Timothy Leary announced his candidacy for governor of California on the station in May. Ponek said in his monthly report the announcement was exclusive but dubious. John Lennon of the Beatles spoke on the air by phone about People's Park and said he hoped a march on the park planned for May 30 would not end up in a riot. Transcripts of Lennon's talk were printed in the *Barb*, the *Good Times*, *Planet* magazine, and the *Stone*. Ponek credited Nisker for bringing both Lennon and Leary onto the station and reported that Roland Young, the station's one radical black disc jockey, hired by Donahue, had talked with the Berkeley police about People's Park one night on his shift and aired rebuttals called in by listeners.

"Oh Happy Day " by the Edwin Hawkins Singers was a top-selling single in May. Voco said he broke it one Saturday night, probably early in April, when he played it once an hour on his show. The following week, the song was aired on the station about once every two hours. The staff liked it, Voco said.

Ponek and Duff talked with each other in May about improving the station's programming and one of their concerns was the familiarity of the music the air staff was playing. Duff said he felt from the start he ran counter to the staff in his thinking on that. He thought they should be less exclusive than they were in what they played. It seemed to him they did not care about getting a new audience and introducing people to the music. They played long cuts and were self-indulgent in their selections. This contrasted with what he thought would be a truly progressive and inevitably commercial style. In a commercial style, a man would play more different kinds of music than they did, one piece shortly after another of a different kind, and so appeal to a more varied and probably larger audience in the same span of time. Pigg, Duff thought, had a good commercial style, but even he did not always exploit it. It was programming theory, Duff thought, to assume the staff would have to develop a more commercial style if they were to grow, and they had to grow. They would have to become less exclusive. Duff said he tried at first to fight them with programming theory.

Ponek could sometimes see the relevance of Duff's thinking. At other times he was not sure. The staff, he agreed, was self-indulgent. It seemed to him they had personalities on the air who were not really personalities yet. They were not professional enough. All the personality they could express was with the music. A guy might play a solid half hour of one artist or one type of music he liked and think that was great. Ponek said he did that himself. He once played a solid half-hour of Weavers and once a half-hour of ragas. Maybe there were 20 people out there who liked it. But then they stopped calling. You could begin to develop conceit about your taste. You would think you knew what you liked and that your preferences were final. Then other people around would say, why not play some of this? If you did, there was an extension. After a while you got to be more of a popular music personality. You could get above the music and select what you wanted. You were out of the jungle. You could go through more changes in an hour than you had been able to before, and more smoothly. Then you could program any station. You had a comercial style.

It took a year, maybe longer. Even then it was easy to lose touch. You would always wonder about the coming week. You wonder if it would be good next week just because it was good this week. In the beginning, you thought you knew what you were doing with your personality. Then later you did. In the spring and summer of 1969, Ponek felt the air staff at KSAN thought they knew, but really they did not.

Duff talked to them a lot about "the thread of the familiar." It was Donahue's line. McClay recalled Donahue started using it during early KSAN. Donahue said he thought of it with respect to top-40 radio and the approach they had of mixing new records with older hits. He talked about it with Duff when Duff first came to KSAN and Duff grabbed it. He grabbed it hard, too hard. He used it to the exclusion of other things. It became a popular music formula with him. Donahue thought both Duff and Ponek tried to do too much with formulas and with rules. When he was program director, he ran the station without setting down a lot of rules. When Ponek took the job, he started trying to lay down rules, and it did not work.

Ponek said he felt he himself was on an ego trip in the spring after he became program director. Part of it was sexual. He was fucking anyone. He almost broke up his marriage, he almsot annihilated himself. Then in the summer the station went through a period when it seemed to him there was a crisis every Monday. He would come to work and find someone had said "fuck you" on the radio. The staff was doing a lot of political rapping on the air between records. People's Park brought it out especially. According to FCC Fairness Doctrine, the station had to be covered for any political attacks made on the air. They had to send transcripts and letters to people, that "on such a date at such time, you were attacked. According to fairness, you are entitled to respond." Duff began developing a file of cover letters. There were also memos to the staff telling them whenever they attacked someone on the air, they had to let the program director know. Ponek felt those memos were covers too, stated as policies. But the staff ignored them or made fun of them.

They also made fun of memos Ponek wrote about the situation of the station. One, in particular, caused him trouble at an early programming meeting.

It began:

Probably every inventor of this decade (with the possible exception of joe and harry Polaroid) has lost the returns of his creative discoveries to less imaginative but shrewder imitators.

KSAN, while smugly proclaiming itself as the expert, the originator, the enlightened radio station is fast approaching the point where its leadership in the field of creative programing and successful radio could go down that same road.

During the period of time when we have been feeling quite happy about our history of success and preposterous imitators, a whole new situation has developed around us. We are in the most competitive market in the country for any kind of radio, and there are 5 count em 5 others in the local FM band who are putting every ounce of their creative or imitative skill into knocking us off. All our competitors rely on our musical tastes to set the pace, but put on their air only the cream of what we play. In short, we are defining the thread of of the familiar, they are playing it, and nothing else.

We have also made the serious mistake of continuing to think that whatever pleases us the most is what is most pleasing to our audience. That was a valid way to approach a program only when it was the only ball game in town.

On June 24, *Look* magazine ran a feature, "The Underground Radio Turn-On," which described underground radio as a growing national trend, quoted Donahue on its beginnings, referred to the KMPX strike, included comments from Ponek on the format change at KSAN, excerpts from Nisker's news, and a statement by Pigg that when the revolution came it was going to be the best-dressed revolution in history. At least a third of the feature dealt with KSAN. The remainder included references to Duncan and WNEW-FM and the fact that profit-conscious owners like Metromedia did not seem afraid of this kind of radio. Advertisers whose products it criticized did not seem afraid of it either.

Yet there was a kind of fear the *Look* article did not mention which could be found during staff meetings at KSAN when Ponek felt like he was somewhere in alley being beaten up. Harris said he thought Duff abandoned Ponek at those times. Pigg thought Duff got his share of being yelled at and torn apart. Ponek got more but he deserved it. The staff, mainly the ones from KMPX, felt it should have been their station and that Metromedia had no right putting Duff in as general manager and Ponek as program director. Duff and Ponek did not know how to run it. The staff felt they did. "We thought we were great," Pigg said, "We did. We all did. It was one of the common things that held us together. We felt we were into something that was really heavy. But it was a struggle all the time to remain heavy. We were conscientious about it."

On July 24, Melvin wrote a letter to his draft board:

Local Board No. 52
Selective Service System
2030 Franklin Street
Oakland, California

Gentlemen:
 I have returned my Selective Service Registration Card and Classification Card to your board. This is my statement of reasons why I have made such a move and I would like it to become a permanent part of my Selective Service record.
 Let me start at the beginning of the process which led me to this decision. I graduated from high school in June of 1960 and enrolled in the University of California at Berkeley in September. The University was commonly referred to as the "Red Beach Head" and the "Little Red School House," especially by my fellow members of the University Masonic Club. I felt it my patriotic duty to help rid the campus of the subversive elements which had given the University such a bad image. So in the fall of 1960, I contacted the Federal Bureau of Investigation in Oakland and volunteered my services as an undercover agent.
 I was employed as a salaried undercover agent for the FBI for almost three years, during which time I joined and informed on several "subversive" people and organizations, such as the W.E.B. Du Bois Club, the Young Socialist Alliance, the Progressive Labor Party and several minor "front groups" like Youth for Jobs. I was a full time "member" of the Communist Party for over a year — all the while serving dutifully as an undercover agent.
 During these three years of dual involvement with United States Government Agencies and the groups on which I was spying I came to the realization that the names and claims of rampant subversion were based on fear and not fact. Among the thirty thousand students on the University campus, there were only fourteen members of the Community Party — hardly a "Red Beach Head." More important was the fact that these so-called "front groups" such as Youth for Jobs really were more concerned with getting employment for people of the ghettos than with recruiting members to the Community Party.... .
 In no way did my work as an undercover agent advance the cause of freedom, justice and equality for the people of the world; it only set it back that much farther.... .
 Peace will not be achieved by violence.

Melvin at the time he turned in his draft card was still working at KSAN. He had quit as a salesman in September but had kept his air shift on Saturday nights. When he turned in his card, he sent copies of the letter he wrote to his draft board to the staff at KSAN. He wanted his friends to know where his heart was. Bear, who was also working weekends, read Melvin's letter on the air. Melvin said he had not asked anyone to play it. Bear's intentions, he felt, were good, but Bear did not do him a favor.

Melvin continued doing his shift for several weekends after Bear read his letter. He was told not to mention his draft refusal on the air because the station could not endorse illegal acts. He thought Duff listened carefully to his shows and that Duff expected him to mention it. But he did not. Then in the end of August he quit. He announced on the air on his last show, "Tonight will be my last show, maybe." Afterward he told Duff it was. Duff wrote in his monthly report:

> Some adverse press is likely over the resignation of Milan Melvin. He was directed to be very circumspect about commenting on his draft resistance posture on the air. Naturally he is saying publically that he was being quashed under the thumb of censorship. However, the parting was very amenable and I do not believe he will actively seek out publicity.

Ponek wrote in his programming report for July there were five competitors to KSAN in the local market. They were in many respects incomparable, but they were able to offer FM rock programming with fewer commercials than KSAN. One was KMPX which had changed its staff again and Ponek said they sounded worse. Another was KGO-FM, an automated station, one of seven in the ABC Love Radio network. There were rumors KGO-FM would go live or use locally produced tapes, but at present their programming was canned from New York. There were two other automated FM rock formats: CBS-FM with its young sound, and KOIT-FM, Mother KOIT, which Ponek said was degenerating. The fifth competitor, KSJO in San Jose, had begun with a loose style like that of KSAN and KMPX but in July was tightening up on program controls, to the point of sterilization, Ponek said.

The station's promotion director arranged in July for a new rate card to be printed up, KSAN Rate Card No. 10, the first advertising rate increase since the group from KMPX came on. It was dated effective August 15.

Bear resigned his weekend air shift in August soon after Melvin resigned his. Bear said he left this time because he was having trouble with Ponek and he had a chance to become involved with a new television show of the Johnny Carson variety. Gleason's special on the history of San Francisco rock music ran on Sunday, August 17. Eldridge Cleaver called from Algiers in August to say hello. Both of the station's cars were taken to the dealer's for check-ups. A Wild West Festival was cancelled, and the news department added a second man who was 21 and black.

Pigg and Ponek went to Woodstock on the KSAN Festival Flight in mid-August and called in reports from the Festival site where there was an attendance of about 500,000. Gleason in his *Chronicle* column of August 20 said the Woodstock Festival was indescribable and remarkable for two things. The first was that there was no trouble caused by the fans who attended. The second was that the chief police officer at Woodstock had received training in handling crowds of young people during his tenure as deputy sheriff of San Mateo County.

Broadcasting magazine of August 11 came out with a special report on "the new respectability of rock." According to the report, the new rock music was different from the old rock'n'roll and was respectable if only because radio stations playing it were running away with the ratings. The report referred to the format's beginnings on KMPX and quoted Duncan of WNEW-FM as saying it had been some time since he last had to reassure a media buyer that listeners to WNEW-FM were not penniless, shoeless hippies, but educated, affluent young adults.

The August 23 issue of *Cashbox* carried an article by Donahue titled "Metanomena" in which he began by saying underground radio was a rotten name, but free form was not quite true either, and progressive rock had little at all to do with it. He said at first he had thought the format they introduced on KMPX would replace top-40 radio. But now he felt that was wrong. It had not replaced top-40, and never would, because top-40 had a basic audience of its own, a younger AM audience. Donahue mentioned that he had left KSAN in April, but he felt the station under its present management was continuing to be unusually effective and an important voice in San Francisco radio. It had changed over the past two years. The sales department, for instance, now had four leading department stores and three brewing companies on the air as regular advertisers, and this was a long way from the head shops and hippy sandal makers who had kept them alive on KMPX.

The San Francisco *Good Times*, previously the *Express-Times*, ran a two-part feature on KSAN in August, written by a member of the *Good Times* staff who had worked for KMPX during the spring of 1969 and applied to KSAN for a part-time job in late July or early August and been turned down. It was after this that he came to the station to get information for his article. He started by saying the interesting thing about KSAN was that it was a station trying, within the context of a heavy business structure, to be a kind of radio everyone wanted to see. The question was, could it work? To find out, he had gone and talked with several of the station's staff.

He spoke with Duff and concluded that Duff was a real businessman, someone with Ayn Rand in his past. Duff took him up to the roof of 211 Sutter Street to talk and when they got into the elevator to go up, Duff pushed the "door close" button, which the *Good Times* reporter interpreted to mean Duff was in a hurry, busy, an executive type. The *Good Times* reporter said he liked Duff. He did not like Ponek. Ponek had turned him down when he applied for a job with the station a few weeks earlier. He mentioned that in his article and said maybe it prejudiced his opinion against Stefan Ponek, but he thought not.

Of all the KSAN disc jockeys he talked with, the *Good Times* reporter liked Young the best. He described Young as a philosopher, a rapper, revolutionary, and a sorcerer, KSAN's Pinball Wizard. He said Young got away with more revolutionary raps than anyone else at the station and that maybe that had to do with his being black. Young played new jazz, did not mind being called a disc-jockeying motherfucker, saw his purpose as relating to his brothers and sisters, and saw himself as a vehicle for pushing the revolutionary struggle. The *Good Times* reporter said he watched Young in the studio at the control board and saw he worked with skill and concern.

Pigg, the *Good Times* reporter felt, came in second to Young in caring about the station. It seemed to him Pigg was the warmest, most affectionate human being at KSAN, and surprisingly unspoiled after having worked for three years at the top-40 station KYA. Pigg was an Aries, a searcher, not an intellectualizer. He would go on dope, then off dope, on brown rice, then off brown rice.

The *Good Times* reporter also talked with Prescott. Prescott told him he felt KSAN was about evolution of individuals, and he felt his job was to make people more aware of themselves and how they related to the social system they lived in. Bear, who was working weekends, said he thought weekends were special and that the weekend man had an important role. But the station was generally careless in its approach to weekends. Harris said he felt the task of the sales department was to make commercials a bridge between the music. Ponek said the politics of the station were touchy, but he was not paranoid about the revolution. His major fear was a military takeover.

The *Good Times* reporter said he asked the staff about money. Stone told him he made enough to get his teeth fixed in a society which did not accept that as a community responsibility. Stone said he knew he made more than 90% of the world's population, but he felt if the money started to entrap him, rather than contribute to his personal development, he would know when to get out. Prescott said he felt money had tempered the concern of the staff. It had affected their belief in what they were doing, shading it from a spiritual belief to a more material one. Nisker, who the *Good Times* reporter said was a real artist and whose coverage of the People's Park was largely responsible for its success, said he felt the staff at KSAN had not really gotten into each other's heads on how they felt about the revolution. As a result they did not have solidarity about it.

PART III:
PROFESSIONALISM

Chapter 8

A VOICE OF THE REVOLUTION
(September-December 1969)

This is a time of search for new standards of conduct which will enable the station to operate successfully in the realm in which it is increasingly implicated. Circumstances surrounding two firings give indication of some of the ways the situation is understood to require new limits. There is, however, uncertainty about the necessity for imposing them and the climate of the station remains remarkably permissive.

Ponek felt the staff hated him. Donahue thought Ponek felt they hated him more than they did, and he thought Ponek misjudged the extent to which the staff was a group. Many of them had come to KSAN from KMPX. They had been on strike together from KMPX. They were like a family. But like a family, Donahue said, they had known each other long enough not to like a lot about each other. Donahue thought Ponek learned a lot about the music from the staff from KMPX, but he did not develop an instinct for it, he was a lightweight among them.

Ponek felt he was the victim of gang war in meetings the staff had at KSAN during the summer and fall of 1969. This feeling reached a peak with him in October. Thinking back on it, he felt that the staff had needed a father figure and he was not cut out for it. Pigg said Ponek was the butt of the staff in their meetings, but Duff got yelled at too and people were generally pissed off. A common thing was someone telling someone else, "You can't tell me what to do." Pigg said after a meeting he might think that was a little hard on Ponek, he does not deserve all this. But he would not say so aloud. He expected if he did, someone else would turn to him and say, of course he does, that son of a bitch.

Bear thought he himself would have made a better program director than Ponek. It seemed to Bear that Ponek was apologetic for what he did, he was not on top of the job and Duff either could not or did not do much to back him up. Prescott felt he could have done better than Ponek as program director and that Duff was always apologizing for Ponek. The subject of staff meetings was usually something Ponek said he wanted to get done: format controls or what records to play. It was hard to know if ideas Ponek would bring up were his or Duff's. Prescott said he felt the meetings were to keep the staff interested in the whole of what happened at the station. Otherwise they might only be interested in private gossip. He thought Ponek tried but the meetings got bogged down in technicalities, like details about use of the elevator. Bear said he felt Ponek almost disappeared sometimes at their meetings. There would be no way of reaching him then.

Ponek's secretary felt the meetings against Ponek climaxed in October and that everybody at the station was on an ego trip. You had to be to put across that personality and make the audience think you were better than them. Campbell felt the staff meetings were distressing during the fall. They were like encounter groups. She would go and stay for part but find herself getting into knots inside and then she would leave. She had work to do. Those meetings, for her, were a waste of time. Duff called one of them explicitly as an encounter group. He said he wanted to get out what was bothering the staff. Street acted up and he ordered her to leave the room. She left, but then he let her come back. Campbell said she thought he should not have let her come back. That was typical of him, he could not fire anyone.

September was Duff's sixth month as general manager at the station. He had the department heads use their sections of the September report to review and evaluate their progress over the time he had been there. That period, since April, he said in his introduction, had a shake-down, slowdown quality. There was considerable turnover and replacement of personnel, and this included the department heads who had to learn new responsibilities and to work in an increasingly stiff competitive situation with respect to other stations.

KSAN's superiority in the ratings was not so overwhelming as to overcome the lower advertising of the other stations. Duff said in his report that KMPX had launched a sales drive which hurt the station in local retail sales during the summer. This was coupled with the summer Pulse showing KMPX doing better than KSAN. Other direct competitors had been gaining numbers during the past six months and were getting a small but significant slice of the record business. The record business, Duff said, was previously exclusively KSAN's.

But the competitive picture had another side. Duff wrote:

> Since there is plenty of money in the market, the more effective sales efforts likely to come from the other stations should pull more total dollars into

FM and specifically FM rock. When I see major agencies buying two FM rockers, I'll know the legitimizing process has begun. Obviously we have to be better and more aggressive than ever to establish our preeminence.

Duff also said in his report that the station's basic programming strategy was to play the right music while maintaining a consistency of sound. The basic sales strategy was to make inroads into conventional or agency-based business, because that was where the mother lode was. But the station had a problem in that it was hard enough to get accounts without, in addition, requesting creative variance so commercials would fit with the format. The solution, he felt, lay in gradualism from two directions: "The deep conviction among our program people that commercial matter must be in keeping with the format will have to flex some, and hopefully the hard line of Metro Radio Sales will soften to whatever degree is practical." In summary, Duff said, he expected there would still be trying times ahead on the P and L, and the station's capacity to educate the agency community, continue to grow in ratings, and the extraordinary competitive picture were the elements that would affect their future.

The promotion director said it was characteristic of Duff in the beginning to think of KSAN as just another radio station. He did not accept it as something special.

Ponek fired Prescott in October. Prescott felt he had differences with Ponek all along, and he could see how his firing in the end of October was one more. But other than that, he did not understand why Ponek had to fire him and why he had to do it the way he did. Ponek fired him while he was away on vacation and replaced him with the Congress of Wonders, a theatrical comedy team. The Congress was already doing the morning show when Prescott got back in the start of November. He found a note from Ponek saying to come see him. It was then Ponek told him what they had done.

It had been announced in a staff meeting while Prescott was away that he would be fired and replaced by the Congress. Street had recorded the meeting. Prescott said he got the tape from her and listened. It seemed to him there was not much response from the staff when his firing was announced. McClay asked why he had to be fired, but the group was more concerned with other things.

Prescott had been one of the first to join Donahue at KMPX. It was in the living room of his apartment on Greenwich Steps that the staff of KMPX had met in March 1968 and voted to go on strike. When the group of them started at KSAN, he was their morning man and their first voice on the air. It was now a year and a half after KMPX and Prescott thought his firing showed what had become of them. They had let the family fall apart.

Bear had been the first of the KMPX staff to be fired from KSAN. That was eight months before, in February. Prescott said he felt Bear's firing then was justified. It was a matter of poor performance, Bear had messed the commercials. His own firing, however, he did not consider justified. He knew the ratings on his show were low but that was what happened to the morning show in this type of format. It followed the others in the ratings. It did not lead. Prescott assumed a decision to fire him and replace him with the Congress of Wonders was made, and made by Ponek and Duff, both Easterners, while he was on vacation . He had no prior reason to expect it. He was shocked when he came back. He was not one to boast but he had felt good about his show before he left. It had variety. He had been doing headline news at 7:30 and his wife had been doing the weather. His news was not without charm.

Prescott did not at the moment remember, but on September 15, Duff had sent a memo about his news to Ponek and Nisker: "Steph/Wes: We can't have Prescott doing the news with editorial-by-innuendo-and-inflection; and with nonsensical sound effects in the background." He did remember from February through April running a "send me your dreams" feature on his morning show. People would write down their dreams and send them in and he would read a few of them on the air. He did an astrology series. He was reading regularly from the I-Ching at 8:00 or 8:30 each morning, a time he thought would be just after the kids had been gotten off to school. He aimed his show at the person at home between 6 a.m. and 10 a.m., the ordinary person, not at freaks or commuters. When he left for his vacation in the end of October, he assumed his show would be covered by one or two of the station's part-time staff. He drew up instructions for whoever would take it and suggested they consider it a contest:

> Dear Cover Cats and/or chick:
> The following is a brief as can 'bout my thing as done mostly.
> 600—"Prescott theme" cart. It ends with the speech of canines, i.e.: "Woof, ralph, arf, ruff, bark," and the like.
> 630—(after ID) "Kiddie's Korner" cart which contains open, close, and ID. The voice is Danny the Kid. Between the open and the close you may insert any recorded bit that suits your mood/that is not music. Spoken Word and Misc. are full of treasures for the filling of this requirement and quality will enhance your Score.

weather
700—I Ching. Best you don't fool around with this wise old bird; or he'll mess up your morning mind. It is my practice to cast the coins after 645 with as open a mind as I can muster. Suggest you use N. Ramani's LP, "Soul of Indian Flute," raga, "Suddha Saveri," as I have been doing for the better part of the past year. Use cart for the opening only. When you finish the reading go right to music. Score for relative advice content of music choice. The curse of too many changes has been known to visit those who are disrespectful toward the wisdom of the ancients.
730—Morning Headlines from front page of Chronicle which you will find stuffed into the downstairs door at a bit before 600. If you come to work early, you will have to make a return trip to Sutter St. in order to complete this assignment. Berio's "Visage" is the chosen background selection using volume for accent. (Unless you already have one.) C.O.W. News Intro cart may aid you in cueing next record. Weather after news. (9361212).
745—Congress of Wonders in their daily reading of someone else's material.
weather
800—R.J. Gleason's "Ad Libs" on Mon/Wed/Fri and Tue/Thur if you want to update column. Be sure to woodshed this item as it is frequently a confusion of typos.
830—News (Tape from previous day) and Weather (Live), weather frequen-
tly whenever freaky. From here on it's music and whatever it is that you do to Score.

Duff said he did not like Prescott's show. He thought the Congress of Wonders could do better. McClay said he thought it was poorly done — firing Prescott while he was on vacation — but on paper it seemed like a wise move and Prescott's program was mediocre. Prescott thought his show was competent, it was not outstanding but it didn't have to be. He thought McClay's show was a similar kind. He had gone to see McClay once at KMPX when he was worried because musically much of the time he felt he was faking it. McClay had told him not to worry. He told him he thought of himself as doing well if he did two or three good shows a week. An attitude like Bear's that each show had to be authentic and a try for the best, Prescott thought was boring. He thought there were maybe three kinds of disc jockeys. There were stars like Bear, workers like himself, and geniuses. The geniuses cared about doing the best show possible and were oblivious to what anyone thought. Ponek felt there was another kind, professionals, and that the Congress of Wonders who replaced Prescott were not professional disc jockeys. They did not raise the morning ratings.

Roland Young in October had been getting good ratings in the evenings and had a professional style. Harris said nonetheless he told Duff, Young would ruin him. It seemed to Harris Young's interest was revolution and not the station and he thought they should have fired Young when Young threatened to resign in late October. But Duff and Ponek were in favor of keeping him on. Harris said Duff told him he thought he could channel Young.

Ponek said he thought he would give Young another chance. He felt he was in part to blame for Young's resignation. It was a case of two cowards. He had sent Young a note critical of his show and Young had responded on the air by saying he could no longer work for KSAN. They had their exchange by note and on the air instead of talking to each other. Ponek had not talked with Prescott before he fired him but he felt differently about that than about his dealing with Young. In Prescott's case, Ponek felt, in retrospect, that he handled it badly — firing Prescott while he was on vacation — but at the time it seemed all right.

Young had gotten angry on Monday, October 27 when he came into the station and found Ponek's note criticizing his show. He felt Ponek was not qualified to judge it. So he resigned on the air Monday night. Then he withdrew his resignation Tuesday and was back on the air Wednesday night. He told a reporter for the Berkeley *Tribe* his leaving the station had been put off for a month or so.

Within a week of Young's resignation, three local underground newspapers came out with stories about it. None had run stories on Prescott's firing. The *Good Times* of October 30 said Young, considered by many KSAN's most controversial and revolutionary figure, announced Monday night that he could no longer work for Metromedia. On Tuesday, the staff of the station discussed Young's resignation in a four-hour meeting previously scheduled as an encounter session. Differences between Ponek and Young, mostly of a personal nature, were raised. Ponek said afterward that although he felt he was the principal target of the session, he was enlightened by it and Young agreed to withdraw his resignation.

The *Tribe* of October 31 said Young, the revolutionary disc jockey, had temporarily reconciled differences between himself and the management of KSAN but that Young said things were changing at the station. The disc jockeys were being told to play more popular music and not to be hard on commercials. He had been told to change the kind of music he played and to make his raps shorter. He had not, however, been told to stop. The *Dock of the Bay* of November 4 said Young quit KSAN on October 27 because the station would not let him use his program to serve the people. The *Dock of the Bay* of November 11 ran a follow-up article on Young's quitting, accompanied by a picture of Young, but focusing more on a description of the station than on details of his resignation. It concluded:

> The contradictions within a system that makes possible both imperialist wars and radical FM stations, both environmental destruction and "good vibrations" are far from being resolved. KSAN's DJ's and managers are part of it, as are we all. I myself have become addicted to KSAN, and consider the news, music and raps an important part of my consciousness. Unless and until there is a viable alternative to KSAN (something like a station run entirely by and for "the people") there is, as Roland says, "no place else to go."

Shortly after 10 p.m. on Wednesday, December 3, Young opened his show on KSAN with remarks transcribed on December 5 as the following:

> Good evening brothers and sisters. This is Roland Young and I'm dedicating this show to David Hilliard.... He got busted today...and to all political prisoners who (garbled) revolutionaries throughout the land...with a suggestion that a caller called in was that uh people who um stand in support of free speech.... For less than one dollar you could send a fifteen word telegram and have it billed to your telephone number....- Send it to Richard Nixon saying "I will kill Richard Nixon or anyone else who stands in the way of our freedom" — as a gesture (garbled). "Seize the time." ("Seize the time" was the introduction to the following song of the same name.)

At about 7 p.m. on Thursday, December 4, two attorneys from the Justice Department and one Secret Service agent visited the station inquiring into Young's program of the night before. Pigg was on the air at the time. He called Ponek who was at home in Sausilito and Duff who was at home in Mill Valley and told them the Feds were there. According to a report subsequently filed with the FCC by Dougherty, Duff spoke with one of the Justice Department attorneys on the phone. Duff then called Young, asked him what had happened the night before, and advised him he probably would not go on the air later that night. Duff then went into the station. Ponek went into the station. The Justice Department attorneys and the Secret Service agent met them there and served Duff with a subpoena. It called for his appearance before a Grand Jury, December 10, and directed him to bring with him various records, documents, transcripts, texts, and tapes relevant to Young's broadcast of December 3.

Sometimes between 8 p.m. and 9 p.m. on December 4, Duff, Ponek, the Justice Department attorneys, and the Secret Service agent sat down together at the station and listened to an air check of portions of Young's Wednesday night show. At about 9 p.m., according to Dougherty's report, Duff suspended Young. Ponek recalled it was about 10 p.m. when Young came into the station to do his show that Duff suspended him. Young later told an interviewer for the *Leviathan*:

> It [the subpoena] was from two federal lawyers from Washington and one secret service agent who hangs around the Bay Area. All three of them came down to the station in a classic intimidating situation. They storm in — I wasn't there when they came in, but I saw them. They're tall cats. One wears a big hat; they have on these grey suits and you know, the greasy look. It really shook up the station manager. Because I used to tell him about those people, but he didn't know they exist. When they converged on the station, that reality was made very apparent to him.
>
> They talked for hours, I would imagine. And when I spoke with him later he said I think last night you went too far. And when he said that I knew exactly what that meant. He meant that I was fired. He said I have to talk to the New York office, the New York attorney, the New York this, the New York that — so in the meantime, I'm going to keep you off the air. And I'll let you know tomorrow. He called me up and said I'm afraid I'm going to have to fire you. And so I said Right On.

It seemed to Ponek that Duff was undecided as to how to treat Young when Young arrived at the station Thursday night to do his show. Ponek went on the air at 10 p.m. in place of Young. He tried, he said, in his own way, to do as revolutionary a show as Young would have done. He had been shaken up by the appearance of the Federal agents. That night of December 4 was the point it came clear to him how the political axe at the station would go.

According to Dougherty's report, the next day at 10 in the morning, Duff called Young at home and notified him he was fired. He sent him a letter on the same date to confirm their phone conversation and forwarded a copy of the letter to the business manager of IBEW Local 202. The letter said Young's dismissal was in accordance with Section VI-3 (ii) of the station's current union contract. Young was being dismissed, as the contract allowed, due to "unsuitability for program requirements."

Also on Friday, Duff arranged for tapes of Young's program of December 3 to be duplicated and made available to a representative of the Federal government and issued a press release concerning the reasons for Young's dismissal. A tape of the release was aired on the station several times during the day:

> Statement by Willis Duff
> Vice President and General Manager
> KSAN
> 5 December 1969
> KSAN has terminated the employment of air personality Roland Young. This action was taken based on the judgment of management that Mr. Young acted in an improper manner during the broadcast of his program the evening of December 3, 1969.
> The specific action that precipitated Mr. Young's dismissal was the reporting on the air of a suggestion telephoned into the station by an anonymous listener that persons in the audience might send telegrams to President Richard Nixon with essentially the same language as was used by Mr. David Hilliard of the Black Panther Party concerning a threat on the life of President Nixon, which resulted in the arrest of Mr. Hilliard.
> It was management's opinion that there was a clear possibility that the statement made by Mr. Young could be construed as advocacy of sending such telegrams, an illegal act, although the transcript of Mr. Young's statement on the air makes it clear that Mr. Young was reporting a suggestion made by a listener only minutes before Mr. Young went on the air. Mr. Young has stated that he does not advocate the sending of such telegrams, nor would he do so himself.
> KSAN is extremely regretful that this event occurred and that there was a possibility of listeners construing any advocacy of an illegal act from one of our broadcasters.

David Hilliard, the member of the Black Panther Party Young quoted on the air on December 3, had been arrested earlier the same day for statements he had made in a speech on November 15 at the Moratorium demonstration at the Polo Grounds in Golden Gate Park attended by an estimated 150,000 people. The KSAN news staff had covered the demonstration and recorded some of the speeches, one of which was Hilliard's. Dougherty said in his report to the FCC that the KSAN news staff, on December 3, had reported Hilliard's arrest at 7:24 p.m. and later in the evening had aired a taped portion of the speech Hilliard had made at the Moratorium demonstration of November 15. They did this, Dougherty said, to place Hilliard's arrest of December 3 in context. The rebroadcast portion of his speech was run on a newscast between 9:30 p.m. and 10 p.m. on December 3, just prior to the listener's calling in his suggestion about killing President Nixon and Young's announcing it on the air.

Duff said there was no question in his mind as to whether or not to fire Young after the Justice Department attorneys and the Secret Service agent arrived at the station the night of December 4 and he found out what Young had done. Young had threatened the President's life. That was a felony. He had done it on the air on KSAN and Duff as general manager was the one they would hold most responsible. Several members of the KSAN staff had the impression it was not so clear in Duff's mind at the time however. Gossett thought Duff waivered in firing Young. But in the end he had to fire him. It did not seem anything else could be done.

Duff said he did not expect the staff to rally behind Young as they did after he fired him. He thought they would be afraid the corporation would come down on them. He himself was less afraid of the corporation than he was of the Federal government. He felt he used the corporation and their legal counsel to come between himself and the Feds. He used them by telling them what he was going to do. He called Dougherty on Decmeber 4 after the station was visited by the Justice Department and the Secret Service. Dougherty said Duff called him and told him, "We've done it again." Duff said Dougherty's response was his usual, "Aw goddamn, Duff." Dougherty said the FBI also called him on December 4 and informed him of the subpoena to Duff.

In Dougherty's mind, there were other than legal considerations involved in determining whether to fire Young. It mattered that Young was black. There were public interest considerations and the black community response to worry about. At the corporate level, Dougherty said he discussed Young's dismissal with the director of personnel, the president of the Radio Division, and the West Coast legal counsel. One of them wanted only to suspend Young, but in the end they agreed on termination due to lack of judgment. You could not be sure, Dougherty said, what Young would do the next time, and you have to be able to trust. You cannot listen all the time.

On December 19, Dougherty submitted to the Complaints and Compliance Division of the FCC an explanation regarding the December 3 broadcast by Young. The Complaints and Compliance Division had notified him by telegram on December 10 of their receipt of a complaint about the broadcast and had requested he submit a detailed explanation and tapes of all matter aired on KSAN between 9 p.m. and 11 p.m. on December 3. Dougherty submitted his explanation in the form of a chronology of events leading up to Young's statement on the air December 3 and his firing December 5. The chronology began with reference to the November 15 Moratorium speech by Hilliard in Golden Gate Park. It proceded with reference to the way Hilliard's remarks appeared on KSAN as news on November 15 and were rebroadcast as part of the news of Hilliard's arrest December 3. It then referred to the visit of Federal government representatives to the station on December 4, the way the station manager Duff cooperated with them and suspended Young, and after consulting with corporate officials, terminated him. It should be noted, Dougherty said, that the station terminated Young despite the fact they were bombarded with complaints opposing this action, although they received no local support for it, and although they were presented with a grievance by the union of which Young was a member. The licensee, Dougherty submitted, had acted responsibly under the circumstances.

Young said he did not think what he did on the air December 3 was all that important. But it did bring things to the surface. He had been feeling for some time that he could not be righteous in that situation. He could not play white rock'n'roll music and silly commercials much longer. The KSAN audience was not his constituency, not the people he wanted to relate to. At one time he thought they were. At one time he had careerist desires to be a disk jockey. Then he got more political and more principled in his taste. The music he liked best was avant-garde jazz in the style of John Coltrane. But to begin to appreciate a musician like Coltrane, you had to listen for a length of time, 10 or 15 minutes at least. You had to be able to follow the runs and accept a sound that was strange. At KSAN the length of cuts they played was normally three to five minutes. They were afraid listeners would tune out on longer cuts.

Young said it was his personal desire not to conform at KSAN and he did not see why, after he was fired, the staff should walk out in solidarity with him. They offered to but he felt they were not really in solidarity with him. They could not be. A walk-out on their part could not have meaning in the context of what was going on.

Boucher thought Young was a clever radio man and he thought Young raised issues that should have been raised at the station. McClay said he backed up Young after he was fired. He offered to go on strike for him. But Young said he did not want the staff to strike because of the possibility of his getting severance pay as a result of arbitration. McClay said he thought Young's quoting Hilliard on the air was probably more threatening in the eyes of the FBI than Hilliard's original statement in Golden Gate Park. Over the air, it might reach more people. The KSAN promotion director said it was at Nisker's house the staff met after Young's firing and drew up a letter to Duff requesting he reinstate

Young. The next day, however, Duff did not come to work and it blew over. The promotion director thought it was like Duff to do that, to avoid issues by seeing if they would blow over, if people would cool, and often they did.

Pigg said he respected Young. He knew Young prepared for his shows. He would get to the station early enough to pull records and listen beforehand to ones he might play. He took his ideas and his music and his approach to doing his show very seriously.

Ponek wrote of Young's firing in his December report: "Repercussions were heavy and members of the black and activist community whose interests were always well served by Roland's show organized write-in protests, petition's, etc."

Newspaper accounts of Young's firing started appearing the day after it happened. Young wrote one himself for the Berkeley *Tribe* and spoke on an interview program on KQED, the public television station. The *Chronicle* carried an article on December 6, "Hilliard 'Support' Costs KSAN Disc Jockey a Job," which quoted Young's statement encouraging listeners to wire President Nixon and said Duff had been subpoenaed for it Thursday night and fired Young on Friday, and that Young attributed his dismissal to a nationwide attack on the Black Panthers and to Federal pressure reflecting the attitude of Nixon and Agnew to tighten up on dissent in the media. The *Chronicle* said Young said he was not himself a Black Panther but that he was a good friend of Hilliard, chief of staff of the Party.

The *Good Times* of December 11 carried an article by someone using the byline, Mead, who claimed to have been in the station the night the Secret Service came to call. The *Tribe* of December 19 carried an article by Young as well as an article instructing people who were pissed off about what happened to Young to let KSAN know, and a copy of a petition they might clip, sign, and send to Duff at the station. The petition said Young's firing was a blatant act of political suppression, an abridgement of free speech, and that KSAN if it was as it purported — an underground radio station attuned to the beliefs and life style of a hip/radical culture — should reinstate him.

The *Tribe* of December 26 ran an article dealing tangentially with Young's dismissal, written by a gay activist who said he had had free speech difficulties of his own at the ABC radio station KGO comparable to those Young had at KSAN. He said he had discussed with Young, who was himself a heterosexual but had no uneasiness talking about or being around a homosexual, the importance of black homosexuals participating in the revolution as such.

After his suspension on December 4, Young consulted a lawyer. Eventually, but not for some months, his case went to arbitration. Young said he pursued the arbitration for the money.

Duff said long before the arbitration proceedings began, he knew in his mind that Young's firing was not just a result of what he did on December 3 but a consequence of a long series of things. These included Young's making political statements in disregard of the Fairness Doctrine, his playing avant-garde jazz repeatedly when he was told to play rock'n'roll, and his downgrading of commercials. In arbitration hearings held in October of 1970, 10 months after Young's dismissal, Duff referred to certain of the incidents he felt had given him cause to reprimand but not yet to fire Young the previous fall. These, he felt, led up to the December firing. With respect to Young's handling of commercials, Duff testified:

> I would say for certain, especially towards the last, Mr. Young stated to me on numerous occasions that although he disagreed with the system involved in the advertising of capitalistic endeavors, he understood that that was part of what made a radio station work, and that there were disadvantages by his own ideology to doing these, and yet the advantages of the job were such that he would do the compromising necessary to deliver the commercials, and such, in acceptable fashion, acceptable being defined as being acceptable to the commercial work, to the clients, and to the servicemen who service them.
>
> However, his performance on the air would contradict that, usually.
>
> Q: After the reassurance to you, you would still have to give him further caution and criticism?
>
> A: Yes. My Sales Manager, Whitney Harris, on numerous occasions told me that he felt that he had to either himself, personally, monitor, roll and show, or have one of the salesmen roll and show so he could be one step ahead, because the guy was going to call the next day.
> The Sales Manager asked me on numerous occasions to discharge Mr. Young, because he was more trouble than he was worth. I thought Mr. Young was an excellent disc jockey, myself, with respect to his ability to communicate himself as a human being, a humorist, his skill in putting on records, so I would resist this urging to dismiss him.
> However, it seemed as time went on Mr. Young's position on these matters became less flexible.

Street took over Young's shift from 10 p.m.-2 a.m. after he was fired. Ponek said he thought she would be good for it. She knew the music and the time slot was right for her, a woman to put you to bed. Ponek thought Duff had reservations about putting a woman on the air. Duff said he had no reservations. Street remembered it was Duff who phoned her and offered her the shift after Young left.

Street was 23. She had grown up in San Francisco, gone to high school in Palo Alto and to San Francisco State College, where she finished and majored in broadcasting. She had wanted while she was there to become an actress in movies or television and had not thought of going into radio until Melvin called her about KMPX. She had worked with Voco and she liked the blues and felt she had also been influenced by Young. In her first few weeks on the air in Young's former time slot, she got a number of hostile phone calls. People called to say she played too much black music or all she wanted to do was suck a black man's cock, but they stopped after a while.

For Ponek, perhaps more than for Street and some of the others at the station, the events of Young's firing carried with them a message that the world lacked a benevolence and a tolerance he once believed it had. So did events of Altamont. On Saturday, December 6, Ponek was one of a group of staff from KSAN who went to the Rolling Stones concert at the Altamont Speedway near Livermore, about an hour and a half's drive from San Francisco. He and the others expected the concert at Altamont would be something like Woodstock. They expected to get high and call in reports to the station as the day went on. They would do interviews at the site and make recordings they could use the following night in a four-hour special, which had already been sold. It did not rain at Altamont on December 6. Someone suggested that might have been why Altamont turned out badly while Woodstock went so well.

Ponek had gone to Woodstock with his wife and son on the KSAN-Diners/Fugazy Charter Flight in mid-August. On the morning of Saturday, December 6, Ponek drove to Altamont anticipating another festival of the Woodstock type. During the week preceding the concert, the station had broacast announcements saying the Stones were going to appear and that the concert would be free. This would be the Stones' last show in a tour of the country they had begun in Los Angeles on November 7. But there were rumors that the concert had been called off, and then that it was on again, and there was uncertainty about the site. The first site considered was Golden Gate Park, then it was the Sears Point Raceway, north of San Francisco. It was not until late in the afternoon on Friday, December 5 that KSAN could announce that the concert was on and at Altamont Speedway near Livermore, not at Sears Point as the morning papers had said. Gleason in his Friday morning column in the *Chronicle* had advised people planning to go to Sears Point to forget their flowers and bring plenty of food and water.

Ponek thought that Saturday morning when he and at least half a dozen of the KSAN staff arrived at Altamont, most of them but himself and Boucher were dosed on acid. It fell to him to do their broadcasts back to the station. He parked his van next to a telephone pole near the stage to get a line in. People had started arriving at the site the night before. By the time the Stones came on, late Saturday afternoon about 5 p.m., an estimated 300,000 people were in the audience. Before the Stones appeared, starting about noon, there were performances by Santana, Crosby, Stills, Nash and Young, and The Jefferson Airplane. The Grateful Dead were there but they did not play. Ponek was on stage when the Stones came on. He remembered having his tape recorder going and occasionally rapping over the sound. But when he got back to the station afterward he found no trace of the Stones on his tapes. The tapes had KFRC on them. His recorder, he assumed, had not been punched to record.

When Ponek started broadcasting back to KSAN from Altamont the morning of December 6, he reported a peaceful gathering and good vibes among the crowd. He said it was as he had expected, and he believed if he talked good vibes it would be more likely to happen that way. Later in the afternoon, the tone of his reports changed. A number of incidents of violence happened near the stage. One was a death by stabbing. The others were cases of people being roughed up. Hell's Angels were involved.

The San Francisco *Examiner-Chronicle* came out on Sunday morning with a front banner headline and a picture of the crowd at Altamont. According to the Examiner's story, 300,000 people had "said it with music" at Altamont. The two mysteries of life — birth and death, were present. Three babies, possibly four, were born at the concert site. Two people died, one drowned in an irrigation canal, the other stabbed to death, knifed twice in the back and once in the face in a scuffle with Hell's Angels. For the most part, however, the *Examiner* said, the gathering was peaceful.

The *Examiner* story on Altamont was said to have been filed before the appearance of the Stones on stage late Saturday afternoon, as if that would account for its generally positive attitude. But a writer for the *Stone* later questioned if the story would have been substantially different if the *Examiner*'s reporter had waited out the concert.

McClay interviewed Jagger briefly at Altamont before the Stones went on. Ponek's van was parked back to back with the van the Stones used.

McClay said he was stoned at the time he reached out his mike to interview Jagger. He saw Jagger had on a shirt with an omega sign on it but did not recognize it for which it was, a Leo. That was how naive he was to astrology then.

Boucher said the KSAN special coverage of Altamont was Ponek's idea and that Ponek was their anchorman on Saturday and he reported a Woodstock. He did not see what was happening. He did not report the terror. It was not until Sunday night when they ran the special that they got it right. Part of the Sunday night special was produced from tapes Boucher and the staff made on Saturday. Part of it was discussion by the staff who had been there, and part was comments by listeners who called in. Three Hell's Angels were among the callers. Boucher said he started getting requests for copies of the call-in part almost immediately afterward. They came from all over. He ended up dubbing and sending out about 50 copies.

Ponek remembered when it came Sunday night he was confused about how to handle the Altamont special. He felt he was not alone among the staff in feeling awkward about having a bad vibes story on their hands and a lot of tapes which were too messed up to use. They finally decided to open the phones and get an audience perspective. Ponek started it by punching on the mike and asking listeners to call in and tell them, "What was it like?" A few of the others took turns with him handling the calls. Their talk was later described as troubled, ambivalent, self-examining, and at moments surprising.

One Hell's Angel called and identified himself as Sonny Barger, president of the Oakland Hell's Angels. He told the station that he and a group of Angels had been hired by the Stones' management to provide security for the Stones during their concert at Altamont. They were given $500 worth of beer as payment for guarding the stage. They had tried to do their job but it turned out badly. In the end, he felt, they had been suckers. Mick Jagger had used them for dupes.

Before Barger called, the KSAN staff had aired a prerecorded statement by the Stones' road manager Sam Cutler. Cutler said the Altamont situation had been confusing, people all acted on their own initiatives, and the Hell's Angels were as helpful as they could be in the circumstances. He did not mention hiring the Angels as security guards to protect the Stones. He said he was not qualified to speak for the Angels about what they did. If the station wanted to know, they would have to talk to the Angels themselves.

When Barger called with his report, Ponek took it cautiously. At one point he asked Barger if he thought what the Angels did while they were guarding the Stones had been worth it:

KSAN: (Ponek)   Sonny, do you think it was worth... .

Barger: Well later, well I ain't no cop. I ain't never gonna pretend to be a cop, and you know what? I didn't go there to police nothing man, they told me if I could sit on the edge of the stage so nobody would climb over me me you know I could drink beer until the show was over. And that's what I went there to do. But you know what? Some cat throws something and bangs my bike or some cat kicks my bike over he's got to fight.

You can say anything you want and call them people flower children and this and that, and there was 300,000 people there approximately, or whatever they say, and I'll guarantee you the largest majority of them were there to have a good time and have a good ball and listen to the people sing and do this and that, but you know what, there were a couple thousand there that was there looking for trouble.

KSAN: (female)   Right. Sonny, just about that question... .

Barger: We were there looking for a good time but you know what, uh, everywhere we go we're looking for a good time but if somebody wants trouble with us, they're gonna get it. We don't want to hurt them people and you know what, they don't want to be hurt but there is some of them lousy people and you can call us lousy people the same way back I don't care. I've been called everything you can be called by experts. But some of them people out there ain't a bit better than what some of the people think of the worst of us man.

KSAN: (Ponek)   Sonny, you got it right there, man, you got it by letting people know exactly where you're at and... .

Barger: Well, you know I'm not no peace creep by any sense of the word but you know what man if a cat don't want to fight with me and don't want to hassle with me you know what? I want to be his friend. If he don't want to be my friend, then outasight, don't talk to me. But if he don't want to be my friend and he's gonna get on my face I'm gonna hurt him or he's gonna hurt me, and you know what? It really doesn't matter if he hurts me. Because I've been hurt before and you know what? I've been hurt by experts but I, uh, over the years I've learned to get up and do it again.

KSAN: OK, thanks a lot, man. I think you've done a lot to enlighten (Ponek) a lot of people as to just what was going on.

A Hell's Angel named Pete called in after Barger and after him one named Andrew. Andrew said Altamont festival officials had asked the Angels to keep people off the stage. The Angels had tried and tried. They asked the people nice but the people did not want to hear it. People near the stage polluted the scene by being high on drugs and one 300-pound man stripped naked and was stepping on people all the way up to the stage.

On Tuesday, December 9, the *Chronicle* ran excerpts from what the Angels said on KSAN Sunday night. The *Chronicle* also ran a report that the Alameda County Sheriff's Department disclosed that the man killed at Altamont while the Stones were on was an 18-year-old black man. He had been stabbed five times in the back and once below the left ear and was seen to have pulled a gun. This occurred while Jagger was singing "Sympathy for the Devil."

In the latter weeks of December, there were discussions about Altamont among the staff at KSAN and in the press in general and they spoke of it as a disaster, as if it marked the end of the era or stood for the way an era had ended, as if somewhere there was being carved an epic in which Altamont had to figure, with the Stones, the Angels, the attending crowd, and KSAN. *Rolling Stone* in its issue of January 21 ran an extensive feature on Altamont. The lead article included excerpts from the KSAN Sunday night broadcast, as had local underground papers in previous weeks. A related article dealt specifically with the positive press coverage which appeared immediately afterward. This the *Stone* referred to as "love generation hype in the news."

Ponek wrote in his programming report for December:

KSAN had the only phone line out of the Rolling Stones' free concert at Altamount Raceway and provided the first honest coverage of the disasterous event, which turned out to be the exact opposite of Woodstock in its flavor. Our four-hour special on the following night was widely hailed as the "finest use of radio in years" by some, and has been widely quoted from.

Ponek resigned as program director in the end of December. He took two weeks off to write Christmas cards, then he called Duff and said he wanted to resign. He remembered Duff told him he was not going to accept that.

Duff wrote in his monthly report for December;

> A severe morale problem developed during the Roland Young affair. The staff petitioned me to reconsider Mr. Young's dismissal. P. D. Stef Ponek demonstrated considerable ineptiness in handling such tricky conditions.
> Mr. Ponek proffered his resignation as P.D. at the end of the month. This was not in protest of the Young dismissal, rather an expression of Mr. Ponek's discomfiture at the responsibilities of the job. I intend to accept the resignation while encouraging Stef to retain some of his administrative duties.

Chapter 9

KSAN
(January-May 1970)

In these four months there is one suspension, one quitting, one firing, and one demonstration in protest of a program. Each raises the question of what will be allowed by the station. Individuals test the tolerance of others involved to determine whether they will stay with the station and how, and their determinations affect the larger course of events.

Pigg said it was on a Sunday night, probably Sunday, January 25, that he was busted for marijuana. The cops came up to the station and got him. The grass was in his car parked in Claude Lane, an alley just across from 211 Sutter Street. They must have followed him, found his car in the alley with identification in it, and then come upstairs to the station. Gossett was on the air at the time.

Pigg said he was in the back room. Gossett ran in just before the cops came in and told him to hide his stuff. He thought Gossett must have been afraid they would bust him for possession in the station. They did not do that but brought him out to his car and said if he would tell them who sold him what they had found in his car they would let him go. Then they took him to jail on the charge of possessing grass and dangerous drugs. He got out on O.R. and the next morning, Monday, called a lawyer. On Tuesday he appeared in court and pleaded not guilty.

Pigg said Duff's first reaction was to take him off the air. Duff gave him a three-week vacation he said was required in order to work out the bureaucratics. He thought Duff handled it well. Donahue said he thought Duff handled it poorly, that Duff did not have to notify Dougherty and New York right away like he did, Pigg had not been found guilty. The station's promotion director said she thought Pigg did not have to tell Duff right away like he did.

Dougherty said he was furious when he heard. When Duff called him in the morning and told him, he thought, "that dumb bastard." He wanted Duff to fire him. It was one more incident of so many. He had not expected them to continue so long. He thought Duff and the staff at KSAN had been told enough, and the issue of someone being on pot, how could they be on the air? It was like with booze. Drugs affected your judgment, and in the radio business you had to be sharp, quick, above and beyond reproach. Dougherty said he thought they had KSAN under control by the end of January. Pigg's bust indicated they did not, and all the trouble the station had been causing him, all the issues he had tempered his reaction to before, converged in one big anger over this one.

Duncan said they had to work out a compromise because Dougherty's immediate response was to have Pigg fired. The president of the Radio Division said they had to calm Dougherty down. They worked it out to give Pigg a paid leave while they gathered information to defend themselves. Duncan said he thought in this business it was not a matter of hiring and firing people but of managing them. He thought the corporation lived up to their responsibility by not firing Pigg.

In an incident seemingly little related to the pot bust of Pigg, the San Francisco *Examiner* on February 18 ran an article titled, "Radio Tips Off Protesters." KSAN, the article said, was giving instructions to Berkeley street people and others staging demonstrations against the trial of the Chicago Seven. The Berkeley *Tribe* of February 13 had advised its readers to stay tuned to the station for information about protest meetings. The *Examiner* reported this and quoted Duff as saying the *Tribe* probably urged its reader to listen to KSAN because it covered the radical movement more thoroughly than any other station. The announcement on KSAN of times and locations of meetings, he said, was not to build attendance but because the station considered this legitimate news.

The broacast the *Examiner* said tipped off protesters had to do with the time and place of a rally held in downtown Berkeley on Monday night, February 16, a rally which was followed by a riot. On Monday night, the *Examiner* said, some 800 young activists who may have been helped by outside agitators, namely the violent Weathermen, but who were for the most part local people organized by a student group, held a rally in Provo Park in Berkeley, after which they staged a rampage through the downtown area and did damage to over 90 businesses, the greatest amount of property damage in Berkeley history.

The *Examiner* article on the role of KSAN appeared on page five next to two other articles reporting on the Monday night events and their possible relation to a national radical terrorist conspiracy. The article on KSAN identified the station as a property of Metromedia, one of three Metromedia stations broadcasting in the area. Metromedia, the *Examiner* said, also dealt in direct mail advertising and billboard display and owned the popular Ice Capades.

On February 18, a California assemblyman sent a copy of the *Examiner* article along with a letter of complaint to the chairman of the FCC:

Dear Dean:
You may have read in the press of another large riot in the streets of Berkeley. Nine policemen were injured — some seriously. This is on the heels of two police department bombings, within the week, in Berkeley and San Francisco.
 I have previously complained that FM station KPFA is operating as a message center; and now this enclosed article states the FM station KSAN is also operating in this capacity. I believe both these stations should be thoroughly investigated because they certainly are lending aid and comfort to the riot leaders. . . .

Dougherty said he felt the *Examiner* article and subsequent complaints to the FCC accused Metromedia of being a voice of the revolution. That, however, was not the cause of Nisker's firing.

Larry Bensky, a new KSAN newsman hired in December, said he felt he should have been fired instead of Nisker because he was the one who fed Nisker the information for the broadcasts promoting the rally of February 16.

Both Dougherty and Bensky, thinking back, referred to what happened as Nisker's firing. But at the time, in the end of February, they said that Nisker quit.

KSAN had been covering the trial of the Chicago Seven and related protest activities in the Bay Area since early February. Bensky had gone to Chicago to work in the Conspiracy office and had been calling back reports to Nisker at the station. Bensky said Tom Hayden called and asked him to come to Chicago to do some work. He asked Duff for a week off and volunteered to phone in stories. Duff and Nisker were delighted. He went to Chicago and worked in the Conspiracy office, helping organize demonstrations nationally to occur the day after the verdict was given in the trial. He got back from Chicago just in time to work on the February 16 rally in Berkeley.

Nisker said he felt he knew what he was doing when he gave the February 16 demonstration in Berkeley a lot of advance publicity. He felt it was important. He produced a program encouraging people to attend. He felt it could be called an incendiary program. He circulated it with a memo to all the KSAN disc jockeys asking them to play it as often as possible on their shifts. But he did not anticipate the rally would turn into a riot. Duff was away at the time the *Examiner* article of February 18 came out. He had gone to Dallas to address a Menswear Retailers' Association convention. Duff liked to do that sort of thing and he thought it might be a way to reach distributors. He came back to the station about a week later. By that time, Nisker said, Metromedia had gotten hold of the article and they let Duff know they were dissatisfied. His firing came down that week, not really his firing, his quitting. Duff gave him the grounds for it by ruling that the station's newscasts would no longer be produced. He also instructed the staff to draw regularly on Metro Radio News for their information.

On February 25, Duff had issued a memo to the news staff:

> Wes, Larry, Glenn:
> In our discussion of the last several days perhaps it has not been completely clear that our newscasts are to be presented without commentary. The device of the produced, satirical newscast will be eliminated for the time being. Please review the memo from David Croninger dated February 13 concerning broadcast language.

Nisker announced that he had resigned from KSAN on a taped news show aired on KPFA, probably on February 27. Bensky said Street announced at the start of her show on KSAN: "For the ten o'clock news, please turn one station to the left to KPFA." An announcement on KPFA said Nisker had quit and his last newscast would be aired on Young's late night show on KPFA after 11 p.m. Young had gone to KPFA after he was fired from KSAN in December. He aired Nisker's last newscast later that night and at the end of it were voices Bensky had recorded of the KSAN disc jockeys each saying their names and saying they used to work for the station: "My name is Dusty Street and I used to work for KSAN. My name is Bob McClay and I used to work for KSAN. My name is Larry Bensky and I used to work for KSAN... ." Then together in a chorus, "This is KSAN, Metromedia 95 in San Francisco-Oakland."

A group of supporters protested Nisker's resignation by picketing outside 211 Sutter Street on Sunday, March 1.

Nisker said he did not hold it against Duff that he had to quit. He got along well with Duff. He felt Duff had protected him at the station and that Duff had a point when he criticized his shows. Duff said he thought Nisker's shows were of decreasing quality for some time before he quit. Nisker said at first when he came to KSAN he had been doing the news in a show-business way, mainly for the theatrics. Then he got into a more political period, a radical political period, and the news seemed to demand more of him. Maybe it was the change in the politics of the time, maybe it was just having to deal with the news day after day, it burnt him out. Duff could understand that, and he was fond of Nisker. At first he had objected to Nisker's wanting to quit. After Nisker left, he wrote and told a listener who complained of missing Nisker that he missed him too and that he had not wanted to see Nisker go, but Nisker was tired and mad at him at the moment.

On Tuesday, March 3, Duff sent a memo to the president of the Radio Division, with copies to Dougherty in Washington and to the corporation's West Coast legal counsel in Los Angeles, stating that KSAN's abandonment of the produced newscast had caused a predictable press, listing appearances by Nisker on television and other radio, and including two articles from the *Examiner* and *Chronicle* about Nisker's quitting.

The *Good Times* of March 5 ran Nisker's quitting KSAN as its lead story. A cartoon on the paper's cover showed Nisker wearing a gas mask and helmet and speaking into a microphone labeled KSAN. Inside was an article written by Nisker:

> The produced newscasts were may way of telling the truth as I saw it. By using music tracks (many thanks to Gene Autry, the Rolling Stones, et al.), the juxtaposition of voices and stories, and lots of splicing tape, I tried to satirize the self-righteous idiots who "govern" the world, to dramatize and give some sense of urgency to the crises at hand, and to provide another perspective, a different objectivity. (Yes, UPI, there is more than one.) When Metromedia decided to suspend the form I had developed they thereby suspended the content and in effect censored my objectivity. Therefore I resigned.
>
> (Fade in the Beatles singing "Everybody's Got Something to Hide Except Me and My Monkey.") It is impossible to communicate with a corporation, even those that specialize in communications like Metromedia, but as I see it Metromedia decided to suspend the produced news for several reasons.
>
> One reason was the fact that I used the barnyard obscenity "Bullshit" in a newscast reporting that Dave Dellinger was jailed for using that very word in a Chicago courtroom....
>
> Another reason was because of our thorough and intense coverage of the Chicago trial....
>
> In general, we must understand that the media controls America by controlling the images which shape the thinking and the desires of the people. The administration knows this fact well. We must get the new images on the air and the screens. The truth about alternative ways of living and thinking, who we are and the truth about our struggle against the corrupt violent death-oriented country we live in. Our truth. Our objectivity. Right

now the media are political prisoners. Free the media! Free the imagination!

I hope to write some more for the Good Times soon — about the media and the movement and other shit. (Close with Delaney and Bonney, "We got to Get Ourselves Together.")

Duff said in his monthly report for February the station had received approximately 400 pieces of mail, many with multiple signatures, "complaining in varying degrees of violence and articulation about our loss of the produced newscasts." Nisker's resignation however, was final. Duff said Nisker had been verging on resigning for some time. The incidents Nisker referred to in his article in the *Good Times* as reasons Metromedia had decided to suspend the station's produced newscasts might be considered part of a case he was making for wanting to leave.

By the end of February, Ponek had had it as operations manager of the station. He resigned from his remaining third of the program director position effective March 1. Duff this time accepted his resignation. Duff said in his introduction to the February report that Ponek had withdrawn from assuming any responsibilities and that Ponek's programming section of the report was a study in withdrawal symptoms. Ponek said in his programming section he would keep his disc jockey shift but Duff would take the program director duties. No one else wanted them. By the end of February, Pigg was back on the air after a three-week vacation following his pot bust. A series of prerecorded lectures by the Zen spiritualist Alan Watts was running on Sunday mornings. Street was still doing the 10 p.m.-2 a.m. shift and liking it. But she felt the staff was into a period of low morale in reaction to the firings of Nisker and Young.

Donahue was back on the air on weekends after a year away. He first came back late in February, or in the very start of March, to fill in for vacancies Ponek said were caused when three of the station's part-time staff were hired over to KMPX full-time. KMPX, in a management change, had brought in Prescott as their new program director and fired the entire staff then on the air to give Prescott a fresh start. Those fired included Larry the Lion, who had been at KMPX since the spring of 1968 when he joined Crosby and Hunt during the KMPX strike.

Of the three who left KSAN to form the new staff under Prescott, two had worked at KMPX in the 1967-1968 Donahnue days. The third had been working at KSAN part-time in news and part-time as a disc jockey since August. A fourth was added in the end of March when the KSAN record librarian went over.

Melvin came back soon after Donahue to do shifts on Saturday and Sunday afternoons. Laughlin also came back to host a listener call-in talk show Sunday nights under the name Travus T. Hipp. Laughlin's first talk show went on the air for four hours on Sunday, March 22. There were technical difficulties having to do with the tape delay unit and speaker phones in the booth, and political difficulties when Laughlin criticized radicals for a recent bombing of a Bank of America building in Santa Barbara. He said he thought bombing the Bank of America was a drag and would not get them anywhere. Then he referred to women as chicks. The *Tribe* of March 27 said a woman called in and asked Laughlin not to use the word chick because its subliminal meaning was about the same as a nigger. His response was to tell her, "Man, I don't want to quit saying chick. Woman is too academic a word. A chick is much warmer and more cuddly. That chick who called up — she'll be liberated when the right man comes along." The *Tribe* went on to say that a demonstration to hang Travus T. Hipp would be held at the time of his next show on March 29 on the fourth floor of 211 Sutter Street. A red bus leaving at 8:30 from the People's Office at 1925 Grove Street in Berkeley would take people over. The protest would be against Hipp's new show and against KSAN for having fired Nisker and Young.

The *Chronicle* of March 30 reported four people had been arrested in a demonstration in front of KSAN on Sunday night, March 29. The *Chronicle* said the immediate target of the demonstration was the station's new talk jockey Travus T. Hipp, but the target more generally was the station's shift away from antiestablishment newscasts and satirical commentary. The *Chronicle* said about 80 radicals, street people, and women militants, most of them from Berkeley, began picketing outside the station's entrance at 211 Sutter Street shortly before 9 p.m. when Hipp's program went on the air. Three of the four arrested were men, two of them were booked for trespassing. The one woman arrested was booked for trespassing and the third man for obstructing a public way. Duff said in his March report he had been in communication with the San Francisco police department during the Sunday night demonstration. He said also he had asked the engineer at KNEW, who supervised the engineer at KSAN, for improved nighttime security in the building: a lockable stairwell door and a lockable restroom door on the fourth floor. The restroom was near the stairwell in the hall outside the front door of the station and he wanted to be sure the nighttime jock would not be locked in or out.

Duff also said in his report that the December-January Pulse had come out and showed KSAN up a full point in all time periods except mornings, and the January-February ARB came out with 100% and 200% increases and a vast jump in female listeners. The January-February ARB, he said, was the first decent ARB the station had had. It still fell short of Pulse and Hooper in its estimate of their share in the market but at least it put them on the map. The February-March Hooper, unlike the Pulse, showed the station with a slight decline overall and an improvement in the morning show.

Duncan became head of the Metromedia Stereo Division in April. This meant he was put in charge of the corporation's six FM stations and the two West Coast AMs, while retaining his position as general manager of WNEW-FM. Duncan said his new position was due to oversight in the corporation with respect to programming.

In the start of May, Duff appointed Boucher as operations manager of the station. Boucher said after Ponek quit as program director and operations manager, and after Duff played program director for a while, Duff put it to him that either he take the position or Duff would get some hotshot from top-40 who knew nothing about the format. Boucher said then he agreed to take the job for $15 extra week and with the condition he could go back to producing commercials if he did not want it. Then the first thing he did as operations manager was give himself a three-week vacation. So in terms of day-to-day functioning, he began in June.

Early in May, Ponek announced on the air the names of winners of 50 pairs of tickets to the premiere showing of the movie "Woodstock," scheduled to open in theaters toward the end of the month. The names of the winners were drawn from a batch of some 400 entries submitted to the station in a contest. The KSAN promotion director said the number of entries was overwhelming considering there had been only four days of on air promotion. When the Woodstock movie opened in Berkeley on May 27, the *Barb* reported, an estimated 250 yippies and friends clogged the entrance to the theater protesting, calling it a media shuck, and encouraging people not to attend.

A "Listeners' Personals" feature began running on the station in May. Duff said in his monthly report, rather than continue to accept phone calls about lost dogs, rides wanted, and the like, the station would now be accepting such informatioon only in written form on postcards and would air it regularly three times a day on the Listeners' Personals feature. Duff said also that the staff had been working with a women's liberation group in May to produce a program of music, poetry, satire, and news which would air early in June and that the ratings were not looking so good. The February-March Pulse showed a drop in female listeners across the board despite an increase in men in the afternoons.

The April-May Hooper, for the third time, showed a drop-off at night and a consequent drop-off overall. The station's chief engineer said in his report a new fire door had been installed in the stairwell to complete office security plans. Chairs, two filing cabinets, and a table were delivered from KNEW-TV, and results of a check of the station's antenna system showed arcing of the horizontal antenna. This, he felt, was an excellent reason to change the antenna system to an RCA circular system.

On Monday, May 4, a friend of Larry Bensky's saw the shootings of four students at Kent State University in Ohio and called Bensky at KSAN. Bensky said it was his biggest scoop and so far as he knew, KSAN was the first in the Bay Area to break the story. At one point he went on the air and announced that the kids at Kent State were shot down in cold blood, not in sniper fire as others were reporting. Duff called him into his office in the afternoon and asked what proof he had. Bensky said he told Duff he knew the guy who was calling in the story and felt he was responsible.

On Friday, May 22, Bensky did a newscast for which Duff not only asked questions but finally fired him. Bensky said about 2 p.m. that day, four kids came to McClay and told him they wanted to go on the air to talk about their just having been fired from Jeans West, a hip-style pants store. They had been working at the branch in North Beach. McClay sent them to Bensky. Bensky said he rapped with them and they told him they had been fired for refusing to take lie-detector tests which Jeans West administered to find out whether their employees were stealing or smoking dope on the job. Bensky said he thought it was outrageous that the tests were given and called the Jeans West main office in Los Angeles but they would not comment. Jeans West at the time had an advertising schedule running on KSAN. Duff was out to lunch. He was still not back when Bensky went on the air with his 3:30 newscast and announced the firing of the four from Jeans West for refusing to take lie-detector tests. At about 4 p.m., Duff called Bensky into his office and told him he was fired. Bensky asked Duff why and Duff said because of a lot of things. Then he got mad and told Duff, "I realized you were a coward and a liar but not a fool too."

There was no 5:30 newscast that day and Bensky said a memo went out saying anyone who referred to the Jeans West firing on the air would get fired from the station. A statement was made late in the day retracting his 3:30 story and saying it would be explained, but it never was. No mention of his firing was ever made on the air. Pigg, however, did say on his show the following night words to the effect: "Larry Bensky will not be with us anymore. It's not my fault. Please don't send in any more letters or phone calls." Bensky said he was crushed that no one at KSAN did anything about his firing. They had with Nisker and with Young. But Bensky did not appreciate the extent to which he was different from Nisker and Young. He was not liked by many of the staff. McClay was an exception. He liked Bensky and he felt Bensky's firing was his fault. The Jeans West people had come to him first and asked him to put them on the air. He had sent them to Bensky. Then Bensky was fired for putting them on. He did not come to Bensky's defense. What he did was give them an excuse for firing Bensky.

Duff said he felt the issue was the defensibility of the station. Bensky kept giving him things he could not defend. Bensky put his own political priorities above those of the station. He editorialized on the air. It was also his personality, his self-righteous radicalism.

Bensky, after his firing, sought recourse in the underground press. He wrote an article for the *Good Times* which appeared a week after he was fired, on May 29, titled, "Ksanitized." He said in the article he had been fired from KSAN after six months because of his failure to follow company directives and policies and that was odd because he had been failing to follow them ever since he began. But maybe it was just as well he was fired. Now that he did not have to go to work at KSAN everyday, he could listen to KMPX. The *Tribe* of June 5 reported Bensky's firing as one more incident now typical of KSAN. You hardly knew it used to be KMPX, what with their ads for Standard Oil, Copper Penny Plastic Restaurants, Luigi's Assembly Line Spaghetti Den, Instant Nutrament, Hamm's beer, Maidenform bras, and the movie, "The Female Animal."

Bensky said the blow he felt after his firing was not only because of the politics, but because he felt he had been doing a good job as a professional newsman. When Duff fired him, Duff claimed the cause was incompetence. That struck a blow also because being fired for incompetence meant he could not get unemployment. He filed for unemployment claiming he had been fired because the job for which he was hired by the station, the job of making produced newscasts, no longer existed. It had been phased out. But Duff fought it. Duff wanted to keep the incompetence claim. Finally, after a few months, Duff gave in. Bensky said he felt in a way he had won. He got the unemployment.

Before he was fired, Bensky had interviewed applicants for the news position left vacant when Nisker quit. One of the applicants he talked with was Dave McQueen. McQueen was from Texas. He was 27. He had a full-bodied network newscaster's voice, a Communist ideology, and radio experience. Bensky was impressed with McQueen and recommended that Duff hire him. Duff was also impressed. He got in touch with

McQueen after he fired Bensky and hired him to start with the station in June.

McQueen said he came by his Communist ideology in the mid-sixties. In 1968-1969, he left his wife and two kids in Texas for about a year and came up to Berkeley. He had worked in radio before in Houston but could not find a radio job he wanted in the Bay Area and ended up working as a union organizer in a chemical company in Richmond. He was in Berkeley in 1969 during the People's Park controversy and he felt it radicalized him. He interviewed with Duff at KSAN but it did not result in a job. Then he went back to Houston and went to work at KPFT until he was fired early in 1970 for inciting a riot. He was fired by the station's general manager who was also his friend. After that he came back up to Berkeley and Duff got in touch with him there.

When McQueen started doing the news on KSAN in June, he had been told about the people who had the position before him. He knew how Nisker had come to quit and that Bensky had treated people as if he were giving them a radical credentials check. He also knew that Duff was tired of having trouble with the news.

Sometime late in May, before McQueen started, Duncan came out and interviewed him at the station. McQueen thought Duncan made a special trip. Duff said Duncan stopped in on one of his regular visits. Duncan was uneasy about the KSAN news. He felt they had inherited a Roland Young situation at the station and it was going to require a professional as a newsman to be fair. He said he told this to McQueen when he met him and he asked McQueen where he stood. McQueen said he was completely radicalized. Duncan then asked how he thought that would affect what he put on the air as news. McQueen said he did not think it would affect it at all.

Chapter 10

KSAN
(June-October 1970)

There is a continuing concern for rules which will distinguish the station's present from its past and which will separate personal preferences, political views, and outside interests from what is appropriate to the station. This concern is to be found in personnel changes and comments by the staff about how they feel they must now behave.

When McQueen began doing the KSAN news in June, Street felt he brought a voice of doom. He would go on the air with his five minutes of so-and-so said today, no smiling KSAN news, and he would not relax on the air. Ponek said McQueen was suspect at first in the eyes of the staff because they thought he might be a narcotics agent and because of the way he sounded. He rang of CBS and network news reporting, old-fashioned professional news reporting, and in the context of that, his radical veiws seemed unreal. They seemed also unreal compared to Nisker. Ponek felt Nisker had the attitude KSAN newsmen had always had, that the news was basically incomprehensible. Nisker would take the logic of events and extend it to the point where it became absurd. McQueen, however, when he found events illogical would get all upset. For him there had to be both reasons and villains.

In the first few months after McQueen started, Ponek said the staff tried to break him. They tried to knock him off balance. The man on the air at newstime would play music behind him or mime his voice, and McQueen would go rigid when that happened. McQueen said he felt it took about six months until they stopped fooling around and accepted him. If he had to date it, he would date it from the time one afternoon when he did a newscast during McClay's show and McClay played stupid music in the background. He let go then and told McClay if McClay did that again he would kill him. After that, McQueen felt, they accepted him.

McQueen said when came to the station, he knew it was a tense situation. On the one side, he could not be radical in any way suggesting the style of Bensky, their previous newsman. Bensky had irritated everyone with his self-righteousness to the point where they would almost rather have dropped the news. On the other hand, and this McQueen said he felt mattered more, the staff was suspicious of him when he started because he was too professional, he was too smooth. His voice was too good and he liked to present things in order. His style was not the one of cultured sloppiness the staff felt comfortable with. Ponek felt when the staff finally accepted McQueen, they accepted that he took his news very seriously, and they came to like him.

Boucher began doing the morning show in June, replacing the Congress of Wonders. The Congress said they were tired of radio. They had done commercials for the station when it was KMPX, and at KSAN they had also done countercommercials, an idea Harris had. They had liked it but radio was not their life's work. They could see maybe doing a comedy show but they were not disc jockeys. They were having trouble with one of their group about the time they stopped and they wanted off by July to go with Donahue on a cross-country trip he was making to film a movie for Warner Brothers. Melvin and Laughlin also left in June to go with Donahue on the movie trip. Stone quit the all-night shift as of June 13, not to go on the movie, but to open a health food store in Oregon. Boucher hired Bear to replace him. Bear's television show had fallen through and he then got involved with Earth People's Park, which also fell through. In June, he was glad to be back.

Pigg was leaving in the end of the month and he was not exactly glad of it. ABC had offered him a lot more money than he was making at KSAN to produce shows for their Love radio network and their local station, KGO-FM. Boucher, in his programming report, referred to Pigg's leaving and the station's other losses of personnel and said June was the month of the big switch.

Pigg said he asked Duff and Donahue what they thought before he quit. ABC had offered to pay him more than three times what he was making at KSAN. Duff told him he could not top that and both Duff and Donahue said not to turn ABC down, but to take the money and run. When the offer came from ABC, Pigg said he was lying down saying to his wife how tired he was of radio. She was patting his head, they were just back together after a split. The phone rang and it was Allen Shaw of ABC with an offer. Shaw had called him once before and he had not known who Shaw was. This time he did and the amount of money, together with how he was feeling, and its being an opportunity to spread what they had started at KMPX across the country, then the encouragement from Donahue and Duff, and he took it.

Pigg was 31. His name outside of radio was Richard J. Quinn. He was born in Sacramento. He had gone to Sacramento State College, been an art major, dropped out, and then went to the Pathfinder School of Radio in Oakland, now defunct, to get his license. His first radio job was on a station in Arizona. The next was in Sacramento for four and a half years, then KYA in San Francisco for three years, then KMPX a few days before the strike, and finally KSAN in the 10 p.m.-2 a.m. slot. When Donahue left, he moved to the 6 p.m.-10 p.m. prime time evening slot. But in February, after he was busted, he felt things began to change for him. He kept having to go to court, there were 15 hearings in his trial, and during that time it seemed to him everything became a mishmash. He had worn his hair long before and sometime in there cut it. He was still into drugs, but tired of looking like a freak and tired of KSAN and of doing two commercials every 15 minutes. He was still trying to do good shows, but there were periods when he did not try. It was hard to get up the energy for it every time. He did things to purify himself like going on a macrobiotic diet and taking drugs. He wanted to be good, really good, to be a saint, but then he would have to turn around and do the commercials.

It would have been different if Pigg had not known what it was to be good on the air, but he felt he did. He felt he knew when he did a good show. It was not the same as doing a show the audience would like. Often he could not tell what they would like. Doing a good show was coming in, and maybe you felt all fucked over before, and maybe you went away feeling lousy, but you got yourself high on the air. It worked like a drug and anyone who felt drugs of some sort were not important to what went on on KSAN was somehow inexplicably beside the point. It was hard enough to understand just what that station was, then to keep up a show night after night five or six times a week. And there was always the expectation that you could do more than just keep it up, you could make it good, real good.

Bear, like Pigg, felt he could tell when a show he did was good and drugs helped. If you were high and feeling fine you could put out good music and make people on the outside feel good. Street felt drugs or not, when she walked into the studio, nobody was going to bring her down, nobody. She was going to do her show and she was not going to be brought down by any program director complaining about the miserable shape of the equipment or any listener phoning in objecting to her music. Pigg said he hated it when people called in to tell him he sounded in a down mood on the air. It was his show and he did not like them personalizing it like that. One reason he used a radio name was to make the show something apart from himself, and he knew, they all knew, there was this implicit rule you should not bring people down on the air. You should not make a show an ego trip for yourself at their expense. Your ego was involved, but you did not want it to become too much that. A lot depended on your musical taste, but that was not just personal. Pigg said he felt he knew what was good, but he learned to play what he did not like because someone in the audience would like it. He kept himself open that way. Sometimes in the first five minutes he would take listeners' calls and play whatever they requested. He felt he was taking a chance doing that, but it was like from God and a way to get out of the ego thing.

When Pigg left at the end of June, Gossett replaced him in the 6 p.m.-10 p.m. shift. Street, when she thought of it was surprised because she felt Gossett was third generation. Gossett was 22, about a year younger than Street but two years behind in terms of seniority with the station. Gossett had gone to work at KMPX in 1968 several months after the strike was over, and had stayed there working full-time as a disc jockey until September 1969 when Ponek hired him over to KSAN for weekends. When Gossett moved into Pigg's spot in June, he felt he had finally arrived. The slot had been Donahue's before it was Pigg's. Gossett had grown up in San Francisco and as a teenager had listened to Donahue on KYA and later to Pigg and looked up to them both. He had gone to San Francisco State College, majored in broadcasting and never finished and in October 1967 taken a job as a producer at KNEW.

He had wanted to work at KMPX in 1968 while Donahue was there and had made an audition tape and submitted it to Donahue, through Pigg, whom he casually knew. But Donahue, after he listened, had told him he did not have enough experience. Gossett was 19 then. He almost worked for KMPX during the strike, he wanted to get into radio so bad, but he felt he knew Donahue and Pigg and he did not want to scab. After the strike when the group from KMPX moved to KSAN, Gossett said he made another audition tape for Donahue. Donahue listened to it and again turned him down and told him he needed more experience. He then went to KMPX and was hired there. That was while KMPX was still owned by Crosby, in the interim before National Science bought it.

Donahue said he liked Gossett when Gossett first came around, he liked rapping with him, but he felt Gossett did not have it together musically. He told him what he thought was wrong, he and Hamilton told him. Donahue felt they trained Gossett during that period of a few years when Gossett kept coming around. When Gossett finally went on the air on KSAN, Donanue felt he understood him and why he played what he did.

In mid-July, an associate editor of *Rolling Stone*, Ben Fong-Torres, also started working at the station, doing a weekend shift. Fong-Torres had covered the KMPX strike for the *Stone* in 1968 and had since developed a role for himself as the one who covered FM for the *Stone*. Fong-Torres was 25. He had been raised in Oakland, had gone to San Francisco State College, worked on the radio station there as an announcer, and been editor of the campus newspaper. In 1967, he took a job as all-night man on KFOG-FM where he played Mantovani music and wrote commercial copy. After that he went to work for the Telephone Company as a magazine editor, did some writing for television, and then in 1969 got the editor's job at the *Stone*. In July, Boucher called him from KSAN to give him some information for an article about the Warner Brothers movie caravan Donahue was organizing. Fong-Torres said Boucher called because of KSAN's tie-in with it. They got to talking and Boucher told him he was in trouble with Melvin going on the movie trip and the weekend people going and asked if Fong-Torres wanted to come on the air and do weekends while they were away.

On July 17, the Friday of the week after he talked with Boucher, Fong-Torres filled in on KSAN for three hours in the afternoon while the staff was over in Marin County on a hilltop having their picture taken. They left him alone with the Kelly girl at the station and although he had been there before, he felt it was frightening. He started working regularly on weekends after that. Duff said in hiring Fong-Torres, they took into account that he was with the *Stone*, but that was not decisive. Fong-Torres had hung around the station, he had a wry style, it was hard not to like him, and they needed staff at the time to replace the ones who were leaving.

Fong-Torres said when he went to work at KSAN, he felt it was important to keep a distance between himself and the rest of the staff in order to keep his objectivity if he had to cover the station for the *Stone*. He felt he made it clear to the staff he did not want to be identified with the DJ commune type of thing they had or go to all their meetings. On the *Stone*, he did not want to cover radio for a while, but because they did not run articles on it too often, that was not a problem. At first he thought of giving the money he earned from KSAN to the Haight-Ashbury Free Clinic, but it was not very much, $20-$25 a week, so he decided he might as well keep it. He felt there were possible legal problems of working in several kinds of rock music activities at once, but he felt Donahue would have an understanding of that and the staff generally kind of understood. But he worried every so often that they might keep him on not because they liked him, but because of his *Rolling Stone* connection. As it worked out, he felt it evolved into a friendship, they liked him and he tried to keep his distance.

Commercial time on the air was sold out in August and Duff referred to this in his report as a new phenomenon. He said it strained the station's traffic, continuity, and billing capacities but everyone adapted and Rate Card No. 12 was formulated for implementation September 1, hopefully to reduce the sold-out posture and to increase earnings.

Gossett said he hated the commercials and he hated it when they piled on more. He felt he could date a change in the number of commercials and in the quantity of what he had to deal with in general as of the summer of 1970. The volume of back-to-school record releases seemed to him worse toward the end of that summer than it had been the winter before. He felt the increase as a pressure and he felt it hit him especially because he was in the prime-time spot and he was new in the spot. Suddenly, toward the end of the summer, he was accountable for a whole lot more. Whereas before he could feel he was mainly educating and exposing the audience by what he played, now it was a matter of keeping up. Groups were putting out more albums and record companies were letting their groups record with other labels. He felt he had to hear the groups live if he wanted to know what was really happening, and maybe he heard a group live and liked them, but they sounded crummy on a record, so he had to check. Then things would change between that album and the next one they put out. He had to keep up with it. He had to be ready if a listener called in with a question. He had to know what to ask of promotion men when they came around. He had to be able to deal with their all-the-time asking, "Have you heard the new . . . ?"

On July 14, Harris sent a memo to the sales staff about the increase in business from rock concerts they were getting. He reiterated cautions in a memo sent by Duncan:

> I believe we all agree that unless the business is placed by a recognized agency, the basic transient nature of concert promoters legislates cash in advance. There may be individual exceptions, but you better make sure you can justify them.
>
> I also question the advisability of any on-air promotions involving these activities — except the concert hall type show. The point is that once we send our listeners to an outdoor festival as part of a station promotion, we must assume some responsibility for the festival being held. Since we cannot do that, I seriously question that we can truly serve our audience by giving away tickets to a festival that because of legal problems on the local level may not come off. Thus, no on-air give-aways, please.

Harris was also worried about the air staff playing around with commercials during the summer. The station's being sold-out meant it was difficult for him to find time to run make-good spots if advertisers complained. On August 31, he left a memo for Boucher which noted, among other things, that Gossett the previous Friday had played a spot for the Gap, the Levis pants outlet, and afterwards called it a credibility gap. On Sunday at 12:30 p.m., Gossett had introduced a Jack-in-the-Box commercial saying Jack-in-the-Box had artificial food. On Sunday between 12 noon and 1 p.m., Gossett had run two spots for Standard Oil in one hour, one at 12:24 p.m. and the other at 12:55 p.m. Harris, in his memo, asked Boucher to remind Gossett and three of the other weekend air staff of the station's policies concerning frequency of commercials per hours, separation of competitive products, triple-spotting, embellishing spots with negative comments, and handling carts, live-overs, and tags according to instructions.

On September 14, the station ran its first fall special, a four-minute program first in a six-week series called "One of Many Possible Futures," originally aired on KLAC. In October, there were four specials of several hours each. The first, on Sunday, October 4, was a quadraphonic broadcast of a concert from Winterland which featured the Grateful Dead, The Jefferson Airplane, and the Quicksilver Messenger Service and was produced in conjunction with the educational radio station, KQED-FM, and the public television station, KQED, which simulcast the concert in color.

Janis Joplin died on October 4 and it was during the Winterland special that night that the news of her death was broadcast on KSAN. It was said to have been first heard as a rumor and then confirmed by *UPI* that she had been found dead in a motel room in Hollywood from an overdose of heroin. The motel room was where she was staying while in Los Angeles for the recording of a new album with her latest band, Full-Tilt Boogie. The *Stone* of October 29 carried several articles about Joplin and on one of them described how the news of her death had reached San Francisco at the Winterland concert:

> No announcement was made from the stage. Concert producer Paul Baratta tried to keep the story from reaching members of The Jefferson Airplane and Quicksliver Messenger Service, who had not yet gone on. All the musicians apparently heard the story as a rumor, but most did not learn that it had been confirmed until after the show.
>
> Out front, the Winterland, which has a legal capacity of 7,500, had sold out at 8:40 and during the evening another 3,000 people were turned away. It was whooping and hollering night, and despite the crowd on the floor and the hindrance of theater seats, there was a fair amount of dancing. About every third time you passed a strong scent of grass, you caught a

strong whiff of wine on somebody's breath. There were a few people calling for reds, but for a San Francisco audience it was a happy, physical crowd, jumping with enthusiasm for the city's favorite bands. Reunions were blossoming everywhere.

Coincidentally, the show was being broadcast live and in color on KQED, the local Public Broadcasting System TV station with quadraphonic sound arrangements with FM stereo radio stations KQED and KSAN.

Whether Janis' death was too unexpected or too hard an idea to grasp, or whether it seemed too remote from the celebratory spirit of the concert, or just because the radio and TV announcers hadn't the class to react to the news (or not mention it at all) — whatever the reason, the KSAN announcers and the hopelessly unprepared man with a mike who led a hand-held TV camera around for KQED handled the news in the worst of taste.

One KSAN announcer, who is underground radio's best-known girl announcer, wandered around backstage asking lame questions about the vibes, where people were from, and so on, man-in-the-street-at-the-Rose-Parade style. She paused during one rap and said, "Wow, man, I hate to bring you down with this, but I just heard that Janis Joplin died in Hollywood. What can you say after that?" Then she returned to the hypnotic repetition of "far out" and "outasite" for a few minutes more, before announcing that KSAN had phoned and confirmed what she had apparently heard only as rumor.

The *Stone* article on how the news reached San Francisco was perhaps unduly harsh in blaming the messengers for the bad news, and in singling out Street, "underground radio's best-known girl announcer," it was particularly nasty. But often they seemed to want more of Street than she gave, and she had not been trained as a newsman to make death wordly. She and some of the others on the staff of KSAN had felt very close to Janis Joplin. In one of the other articles in the *Stone*, Fong-Torres had an interview with Nick Gravenites who recalled a time he had taken Joplin to see Donahue in 1969 after she broke with Big Brother:

> Janis at the time was really upset. She was saying, "Fifty people help Joe Cocker and Leon Russell helps Joe Cocker and all these people help Joe Cocker and all these people are helping these people and why doesn't anybody help me?" She was upset. So essentially that's what I was doing at that time. I would chase around — I'd go over to Tom Donahue's, and I would say, Tom, Janis feels let down. "How is it that certain people can command a lot of fellowship and respect and help from other people and I can't?" And then Tom would go through every one of his old records that he loved and then he'd go over there and he'd sit around and say, "Janis, here's a tune maybe you'd like on this album, here's something maybe

you'd like on that." And she loved 'em, man. She picked up maybe five or six tunes off Donahue.

Donahue was away in England following his movie-making trip for Warner Brothers when Joplin died. McQueen, in his monthly report for September, said the station's news department had started receiving daily news dispatches from *Earth Magazine*'s *Earth News Service*, an alternative news source, and he expected soon to do away with the 6 a.m. and 7 a.m. feeds from Metro Radio News and do the morning newscasts live, using *UPI* copy plus *Earth News* items.

When McQueen started in June, the news department was subscribing to *UPI*, *Liberation News Service*, the New York *Times*, the San Francisco *Chronicle*, and *Newsweek*. They also had Metro Radio News and various local sources, including people who just called in. Shortly after he started, McQueen said he added two more newspapers, the Manchester *Guardian* and *Le Monde*, and dropped *Liberation News Service*. *Liberation News Service*, he felt, was mainly rhetorical bullshit, it cost $20 a month, and he could pick it up out of any underground newspaper.

Bensky, had he known about McQueen's cancelling the station's subscription to *Liberation News*, probably would have thought it in keeping. It seemed to Bensky, McQueen was impatient with the community and with left-wing political institutions and that McQueen's attitude generally was "fuck 'em because they're not together." But McQueen did not see himself that way. He felt up until sometime in November, he was narrow in his focus. He dealt mainly with issues of concern to the radical community and talked to people like himself. Then in November, he had a difficult time with his conscience on one particular story and that set him to thinking about his involvement with radical causes and he decided to change his approach and not go that way so much anymore. He felt it was a deliberate decision he made then to expand his news coverage to reach a broader audience. A broader coverage was probably what the audience he was already reaching wanted, and he felt if he did not expand, he would end up sooner or later just talking to himself.

The KSAN news in October included reports on the arrest of Angela Davis in a New York City motel on charges of being an accomplice to murder and kidnapping in the Marin County Courthouse shooting of August 7. There was coverage of prison situations and the Los Siete trial, an in-depth story on military jobs in the Bay Area and how a Vietnam phase out was leading to an economic crisis in jobs, and an in-depth story on the crime control act of 1970 and how it would affect organized crime little if at all but would virtually abolish the Bill of Rights. There was also one story McQueen later summarized as "The seizure of KMPX by the staff, who broadcast revolutionary slogans until management pulled the plug at the transmitter."

Boucher, in his October report, said: "Our chief competition KMPX went off the air October 21 and is expected to return with a new staff November 13. I'm afraid we just ran them into the ground, didn't mean to lean on them so hard." Harris, in his sales report, said after KMPX went off the air, KSAN picked up some of their advertisers but most were reluctant to pay the higher KSAN rates. They had a $5 rate on KMPX. The KSAN promotion director, in her report, said after the KMPX air staff was fired, KMPX started playing the ocean and would continue with that until November 13. As of yet they had done no outside promotion.

Any listener could hear, at 106.9, KMPX was playing the ocean, sounds of the surf from the far side of San Francisco, day and night for three weeks in late October and early November. The ocean went on after serveral days in mid-October during which KMPX was off the air entirely. The firing of the staff, which Boucher and the promotion director referred to in their reports, was the firing of a Collective, a group not unselfconsciously reminiscent of the staff Donahue had drawn together in 1967-1968. Like the earlier staff, the KMPX Collective of October 1970 claimed they wanted to be revolutionary, to make KMPX a people's station, and to serve the larger community while developing community among themselves. Like the earlier staff, they ran up against interference with the ownership of the station, now the National Science Network which had bought it from Crosby in November 1969, and like the earlier staff, when push came to shove, the Collective first voiced their grievances on the air and then took them out onto the streets and received a rush of publicity.

There were differences however. The year of the Collective was 1970, the enemy was no longer Crosby, and one of the media reporting the story was KSAN. On Wednesday, October 21, KSAN aired a 24-minute newscast produced by McQueen in an Edward R. Murrow documentary style. It began with McQueen:

> This is KSAN News. Last night, KMPX radio went off the air in sudden and dramatic fashion. The management of KMPX says that KMPX will be back on the air shortly with a new staff of air personalities. Today KMPX was on the air, on and off, but without any new personalities. The staff of KMPX went on strike of a sort and this is the story of what happened.

McQueen next referred back to the first rock staff on KMPX in 1967, to their strike, and their subsequent move to KSAN. He went on to describe how certain members of the present staff of KMPX had been asked to resign Tuesday morning and had then locked themselves into the KMPX studios at 495 Beach Street for four hours late Tuesday afternoon. They had talked on the air about their dispute with their management, the National Science Network, until 7:50 p.m. when a representative of National Science pulled the plug out at the transmitter on Mt. Beacon in Marin County and shut the station off.

McQueen's newscast included excerpts from statements made on KMPX by members of the Collective before they were shut off. Their principal spokesman was Roland Young. Young had gone to work at KMPX as a full-time disc jockey in June, leaving KPFA where he had gone after he was fired from KSAN the previous December. Another spokesman for the Collective was Bensky who had been fired from KSAN in May, gone East in June, and when he returned in September taken a job with KMPX. Bensky said when he came back to the Bay Area, it was obvious KMPX was there the action was. It seemed to him the KMPX Collective was trying to make KMPX what it was originally supposed to be and what KSAN clearly was not.

Prescott was no longer program director when the firings and lock-in of the KMPX staff occurred in October. He had left in September before the Collective really got going. National Science had appointed a new program director to replace him and the new program director's exercise of authority was one of the issues around which the Collective organized.

McQueen's KSAN newscast of October 21 included replays of statements made by Young on the air on KMPX while he and other members of the Collective were locked inside the Beach Street studios. It also included statements made by Young after the juice was turned off when he and the others walked out of the studios and spoke to a crowd of about a 100 people assembled outside. On the air, Young had begun by saying:

Young: This is KMPX in San Francisco.

Music: free the people, freedom now...free the people....

Young: All power to the people. Today, October the twentieth 1970, Sally Moses, Jim Mayer, Jon Fox, and Roland Young, the KMPX People's representatives, met with the National Science Network Corporation's representatives. The meeting was designed to resolve some of the contradictions that have existed over the past few months, the contradictions that stem from the air staff's forming of a collective and management's desire to invoke absolute authority upon a corrective, creative, and sensitive group of righteous sisters and brothers.

From the meeting we learned of three directions the management intends to move in: number one that Tom Trunell the recently hired program director will have and invoke absolute authority over all matters concerning the sound of the station.... . They would like to end the so-called radical political direction of the station that started, they said, with the hiring of Roland Young. That means no political interviews, no political raps, and even no political musical programming, that is to say putting two or three records together to say something above and beyond just get up, get stoned, fall out, and act crazy.

Number two, the management would like to hire new disc jockeys with totally different programming attitudes.... . And number three, we learned that they would like Jack Ellis, Joshua, Jon Fox, and Roland Young to resign. We were informed that it is not our air performance that they necessarily object to but rather instead it is our political ideas and organizing approaches. They think that it is the four of us who are responsible for stirring up the trouble and organizing the people in the Collective when it was the Collective that organized itself. We say hell no to all of this. Hell no. We would not accept it and we would not go along.

There followed a chorus of voices, music, and statements by other members of the staff, including Bensky:

Bensky: Another stage has been reached beyond the Mickey Mouse that has been going on with radio stations and the so-called new culture for the last three or four years in the Bay Area, that as Roland just said this marks a turning point. There's no more going back and saying well welcome back, what about one poor job, and put me on the air, maybe I'll listen to your stuff, maybe I won't, sure I'll play some Creedence and not some Coltrane, you know, and that kind of dickering and negotiation. I think anybody that goes on the air on a station like this after we leave should definitely be labeled by the community as an out and out pig.

Voices:	Right on. Right on.
Bensky:	He's not only taken the bread out of our mouths, because don't forget 10 or 12 people are going to be unemployed after today, and that's very serious because he'll also be in effect blacklisted in the Bay Area. Ten or 12 people are going to be unemployed but also the community is losing a voice. And don't let 'em think that if they're putting some little thing on for you every once in a while and saying well sure you're having a damn good performance for a clean the beach program, sure come on down we'll make a 60-second spot for you. Don't let them delude you into the fact of thinking that that's public service.
	Public service is serving not only people who are anxious to clean up beaches but people who are anxious to get themselves out of jails, people who are anxious to get themselves out of schools where they don't belong, people who are anxious to get themselves out of jobs that they can't respect and work that they don't dig doing, that's wasting their lives and wasting their love and wasting everything else.
	So I think that the way that people relate to radio should be by strident demands. We saw a little of this in the women's movement last spring when they came up here and to KSAN and got what they wanted, by strident demands, not negotiating, and by clearly recognizing the enemy as Bank of America radio, vaginal deodorant radio, and all that for exactly what it is.

McQueen, in his newscast, next aired a statement by the management of KMPX, two-thirds of which had been read on the air on KMPX before the transmitter was cut off Tuesday night. The last third had been called into KSAN afterward by the KMPX station manager who asked that KSAN broadcast it. The statement included the following:

Management:	KMPX-FM and its ownership, National Science Network, have resolved an insoluble political impasse by interrupting normal programming and, over the next several weeks, completely revamping broadcast equipment and format....
	KMPX's problems were not immense on-air violations, nor violent disagreements on open programming, nor on the internal functioning of the air staff Collective. The complete and only grounds of conflict between Collective and management

lay in differing concepts of authority, control, and audience structure....

Under different circumstances, perhaps one where workers do, in fact, own and control broadcast equipment, a revolutionary collective might maintain. In commercial radio, where a corporate entity owns and operates a station for fair profit and for benefit of the general public, such power/control revolutionary demands are impossible and harm the corporation, the public and the revolutionaries....

KMPX is supending normal broadcasting.... To those inconvenienced by this action, might KMPX suggest listening to KSAN, KJAZ, KQED (TV and FM) and KPFA and B. And write or phone in suggestions to KMPX at 771-8503.

Future programming of KMPX, with some new air staff, will strive to keep open all the lines of community service and response, including those to the radical element.

If it was not surprising, it was perhaps curious that the KMPX management statement justified the dismissal of employees with reference to a Marxist political reality. Crosby had not done that when he refused to take back the striking staff in the spring of 1968. He had made his case in personal terms reflecting the situation as he felt it. They were holding a gun to his head, he said, and taking them back would be committing suicide. Duff, when he fired Young from KSAN in December 1969, had justified his move with reference to corporate management responsibility in the eyes of the law. Nonetheless, it was possible they were all saying much the same thing, even Young. McQueen's newscast carried some final remarks made by Young when he came out from the KMPX studios Tuesday night and spoke to the waiting crowd:

Young: It's ended as far as this station is concerned, and we have proven that you cannot do it in the context of a capitalist radio station. They got some young white rock and roll boys and they ready to bring them in right now. They got them dressed, they got their glasses on, their hair cut.

But if they were all saying much the same thing, it was not clear why McQueen, in his newscast of October 21, in 10 instances blipped out shits and fucks and other profanities of the sort, and it was not clear why although in his tone of voice he seemed to want to separate KMPX from the world of KSAN, putting it somewhere in a realm of high jinks, he ended his newscast with a comment on the scene outside the Beach Street studios after Young left: "The crowd went home. We went home. Obviously a lot of soul searching will have to be done by a lot of people in so-called hip radio in this city."

It could have been just that McQueen took his news very seriously, but KMPX was a measuring rod as well as a plaything. On October 21, the *Chronicle* reported that members of the KSAN staff had been present among the crowd of sympathizers outside the KMPX studios Tuesday night. Also on October 21, Duff left a memo for McQueen: "I want to be on record that your action in carrying a sign stating 'KSAN staff supports KMPX staff — Right On!' while covering a news event relating to KMPX staff demonstrated poor judgment on your part."

Chapter 11

A DIFFERENT STATION
(November 1970-May 1971)

There are during these few months no dramatic firings of staff and no arrests. Events of the period indicate an ability to deal with an increasingly varied outside world. The nature of relationships with that world is suggested in a request to the National Guard, a response to an oil spill, negotiations over a union contract, news coverage of political events, and support for a movie-making caravan. Although these events are of very different kinds, it matters that decisions about them are of a piece and made with reference to a present which is seen as different from what it was even six months before.

On December 9, Duff sent a letter to Lieutenant Elwood McCann of the U.S. National Guard requesting that Dunlop be allowed to grow sideburns and slightly longer hair:

Dear Lt. McCann:
Doug Dunlop has been an account executive for KSAN for two and a half years and has proven to be very valuable to us. In that time he has risen from a trainee position to senior account executive and is top producer of a staff of four advertising salesmen.
He has been in the National Guard all this time and his attempt to conform with military style haircuts has seriously affected his image as a representative of a radio station which specializes in avant-garde music. He has manfully withstood frequent wisecracks about "not being with it" because of his short hair and clean shaven, youthful face.
In these times when even lawyers, doctors, and bankers are sporting long sideburns, Doug's image has become less and less tolerable as a KSAN representative. I believe that his record of achievement, while good so far, would have been even greater if he had been permitted to modernize his image somewhat. I believe that in the future his credibility in the radio business will suffer even more as neatly trimmed beards, handlebar moustaches and long sideburns gain more respectability in the highest management and professional circles.
I am prompted to write to you at this time to request your permission for Doug Dunlop to grow sideburns and slightly longer hair because I read that Admiral Zumwalt recently approved full beards, long hair and sideburns

for full-time naval military personnel. Since Doug Dunlop is only in the military in a part-time capacity, and his full-time work is with us, I earnestly request your approval of a slight relaxation of the cusomtary military short hair in his case. I assure you it will be very valuable to his career.
Thank you for your consideration.
Sincerely,
Willis Duff

It was not Dunlop but Jeff Nemerovski who brought the first waterbed account onto the station in November. Nemerovski was now the station's youngest salesman. He was youngest in age, 22, and youngest in time with the station. He had started as a volunteer in June after meeting with Harris and telling him about his theory of radio, about the development of commercial stations in college towns and how networks and syndications were the answer and the appropriate Freudian thinking was that antennas in this day and age were like totems but they did not represent the supernatural.

In November, a childhood friend of Nemerovski's from Chicago got in touch with him at the station and said he was into the waterbed business. He said he was starting a company and wanted to arrange to have credit for time on the air. Then one day when Duncan was out for a visit, Nemerovski walked into the studio and asked Duncan and Duff how they would like a waterbed account. Duncan said "yeah, waterbeds." Duff said "yeah." Duff ended up making up the name for the company, Magic Mountain Waterbeds, and helping them by giving them time on the air. Magic Mountain was on for about two and half months, until sometime in February when they went bankrupt. They went out of business owing beds to about 1,400 people and owing $1,800 to KSAN for advertising time. Their billing had gotten up to $900 a week. Nemerovski said when that happened, he felt responsible and the station tried to do what they could to get Magic Mountain to cover the customer orders. At the time Magic Mountain went bankrupt, they had been selling beds two for the price of one at $40. His friend who had started the company had bought a lion for promotion purposes. Nemerovski said he told his friend, "Come on, you can buy a lion but you can't pay us." His friend had formed Magic Mountain into White Tiger Waterbeds in February before it folded, which was why the lion.

Nemerovski said what kept KSAN going in January and February during the post-Christmas slump was waterbed business. Harris wrote in his December report that Nemerovski made the most of the waterbed craze. By the end of December, there were four waterbed companies on the air. The promotion director said in her report the station ran a waterbed giveaway Christmas contest for advertising agency and client friends. Entrants were asked to write a paragraph on why they wanted a bed and winners were selected by a panel of three on the basis of originality and imaginative answer. A second waterbed giveaway for listeners ran January 23-February 1. In that one, the station gave away 20 beds, two a day, with disc jockeys selecting winners' names at random from a pool of 22,000 entries. The beds were contributed by companies who had advertising on the air.

A single waterbed was installed in the lobby of the station in January and sometime around then everyone on the staff got one. Pam Sanders, the traffic director, said first the disc jockeys got them, then the others got them. All it was was a bag. For it to last, you also needed a liner, a frame, and a pad or heater. More waterbed companies went on the air after Christmas. By the end of March, the names of companies who had been on included: Magic Mountain Waterbeds, Waterbeds Unlimited, Neptune Waterbeds, White Tiger Waterbeds, Pasha Pillow, Embryo Waterbeds, Waterbed and Company, Waterbed Factory, Environmental Valve, Porpoise Mouth Waterbeds, and Undulator Waterbeds. Then it seemed suddenly they were all off. In his monthly report for May, Harris said Gardner had the last waterbed account remaining on radio.

Mike Lavin, the owner and manager of Undulator Waterbeds in Berkeley, said what happened was in the spring of 1971, all of the companies burnt out, his own included. He had started Undulator before Magic Mountain in the fall, but Magic Mountain started advertising on KSAN before him. When he went on the air himself in December, the weekend following his first week on, 125 people came in the front door. Then in March, after Magic Mountain went out selling two beds for $40 and Neptune came on with one bed for $25 and a second for $15, he went off the air.

Stan Weinberger, who owned Mr. Broadway, was also the owner of Neptune. Weinberger said he started with waterbeds after Magic Mountain. He went on the air on KSAN at first in one of their giveaway contests. Then it became a bandwagon business. The fabricators were the problem. So far as he knew, waterbeds started in San Francisco and KSAN was the main one responsible for promotion then. It was not until six months later that they started in New York. Lavin of Undulator said the first waterbed store opened in July 1970 in Culver City, but San Francisco was the best market.

A new promotion director, Joanne Dymond, joined the station in November. She was 32 and she thought of herself as part of the professional art world. She had not worked in radio before. She had once been a painter and had graduated from San Francisco State College with a degree in art. She had been in business on her own doing public relations for artists and galleries and for the San Francisco Erotic Film Festival. She said she applied for the KSAN job when she heard about it. KSAN was her favorite station. From what she knew, Duff interviewed about 50 people for the job before settling on her. All of them were women, they came cheaper. He was interested in finding someone who could upgrade the press the station was generating and the advertisements, and who could work in two directions — with the undergrounds and with the establishment press.

Among the first things Dymond did when she started in November was to expand the station's "What's Happening" feature to include art exhibits and films. She revised and expanded the press list. She moved the press release production from two locations to one and had the addresses for mailings put on magnetic tape instead of addressograph, and she tried to involve the staff more in putting out press release material. But in one meeting in January, when she proposed they get out a newsletter, they turned on her. They said she would ruin the station, that she did not understand what it was to be in the counterculture. Their leaders, she thought, were Laughlin, Ponek, Bear, and Street. So she dropped the newsletter idea. Once for an ad in *Coast* magazine, her first photographic ad, she had in mind posing everybody on a waterbed. They attacked her for that. She felt they devastated her. They said she was trying to undermine their credibility. The ad ended up with a waterbed and someone on it, but not anyone from the staff.

Dymond said she felt the staff had a general distrust of her from the start, and it did not have to do with what she proposed. It had to do with her being too New York and not smoking dope or being funky enough or easy with them. She was too professional. And that was too bad because in some ways she liked them and she could see many possibilities for promotion in them. But she resented their ego tripping, their star psychologies. When they were up, they were up, but when they were down they were impossible.

Campbell felt Dymond was too advanced for the station and that she was out for herself and her own professional advancement more than she was for the station. But Duff liked her and gave her her head. Duff said he felt Dymond had class but not too high-class. She could generate press. But he knew the staff did not like her. McQueen said he liked her. Dymond became promotion director full-time in April and Duff got someone else to be his secretary full-time. Up until then, one person did both jobs.

Duff felt he was in some ways riding high in the winter and spring of 1971 and there were things that happened at the station that he loved. One was an oil spill cleanup which occurred after two Standard Oil tankers crashed under the Golden Gate Bridge early in the morning of Monday, January 18. The spill left 80,000 gallons of oil and many dirtied beaches and dying birds around the Bay and a slick that extended along the coast from Bolinas to Pacifica. On Tuesday morning, the day after the tankers crashed, an oil spill cleanup campaign went on the air on KSAN. A volunteer rescue organization set up headquarters at the station, using office space, phones, the station's credit rating, and air time to coordinate efforts of people who wanted to help clean up the mess and save the birds. Among its activities on Tuesday, the volunteer group got out several thousand copies of a printed leaflet with "vital information" about care of birds. It began:

Caution!!
Birds will be hysterical. Many will be blinded and having difficulty with breathing as well as being unable to fly. A person approaching will further frighten the bird to the point where it will claw, flap, bite, peck, or scratch you. Wear gloves and heavy clothing, and treat any cut promptly as it will become infected. Remember that pelicans, cormorants, gulls and herons are large and strong birds and must be handled firmly to avoid injury to yourself or to the bird.
    The first thing to do with an oil-soaked bird is to get it warm. By wrapping it in a towel, the bird's own heat will warm it up, even if it is still in the open on the beach. If at all possible, soak the bird with light mineral or cooking oil before wrapping to keep the heavy oil from solidifying. Once the bird is securely wrapped in a cloth or towel with only its head exposed, clean oil away from the nostril holes and the eyes, using a soft cloth or Q-tip soaked in light oil. After the eyes have been cleaned, you can place a drop of people-type eye drops (I.E. Murine, Visine, etc.) in each eye.
    Ideally, you should not try to clean the birds' bodies until they are thoroughly warm.... .

Duff in his monthly report referred to the event as the Great Oil Spill and said that the volunteer group working at the station at its peak numbered about 30 people and became the central clearinghouse for information and instructions for volunteers all around the Bay. He ordered four extra phones for the group to use and gave them a room of their own, what had previously been the office of the station's chief engineer. On the whole, he felt, it was excellent community involvement. A great deal of press was generated. Boucher said in his programming report that within hours of the spill, KSAN became a focal point of the cleanup effort. The first week, much air time was devoted to it. This was followed by a second week of gradually lessened activity and commendation for the station's efforts coming from Washington, local ecology and conservation groups, and most importantly from listeners. And all this was during an ARB rating period.

Paul Lambert, one of the organizers of the rescue group, said he was standing in his kitchen at home making noodles and listening to KSAN on Monday night, January 18 and was so affected by McQueen's coverage of the oil spill that he determined to do something about it. He talked to some friends. They said let Standard Oil clean it up, they spilled it. A little later he heard on the radio that KSAN was giving out phone numbers for people to call who wanted to help save the birds and wildlife hurt by the spill. Heliotrope was supposed to be coordinating, but he knew from a friend they were not. He tried to call KSAN to tell them their number was wrong but he was not able to get through. He then went over to the station. He got there about 8 p.m. and the door was locked so he went across the street to a hotel and called. The number he had was the station's business phone which was not usually answered after hours, but Bear was there at the station and for some reason he picked it up. Lambert said then he went upstairs to the station and explained to Bear how he felt somebody had to serve as a communications center for San Francisco. Other cities were organizing. Bear told him the regular staff at the station was not going to do it, did he want to do it? Lambert said then he and a friend of his named Peter started answering the phones.

At first they thought they would just be getting information out, but it became obvious as they answered the phones that people needed supplies. It became known what they were doing and supplies started coming. About 10 p.m. or 11 p.m. that night, people started bringing supplies into the station. But it was clear they could not keep them at the station. They sent them out, by way of a call, to a Clementina Street supply house. Lambert said he felt the most real and significant work of the oil spill group was in the first 12 hours. McQueen came back to the station about 11 p.m. People had already been out on the beaches working that day. The *Chronicle* had played down the story. They did not play it up until Wednesday, two days after the spill.

Lambert's friend Peter took over organizing the supplies and transportation. He ended up organizing transportation for people all over the Bay. Peter was 23 and a native of San Francisco. Lambert was 26 and from the East. He had applied for a job as a salesman at KSAN in October 1969. He had been turned on by the underground character of the station and he told Harris that, but Harris said no, he did not want any of the underground crap, he wanted big accounts.

Lambert said a lot of the volunteers who got involved in the cleanup were native San Franciscans and their education level was high, mostly post-high school. Sometime late Monday night or early Tuesday morning, they decided the most significant thing they could do was to get people out information about how to clean birds. Someone volunteered paper. A friend of Lambert's who was a printer and had done printing for the peace movement started printing up their leaflet on the care of birds using his own paper in the middle of the night. Tuesday morning, some Boy Scouts from Hayward or somewhere like that came to get the leaflets and give them out.

On Tuesday morning, when Duff came into the station, it seemed to Lambert his immediate reaction was positive, as long as the station kept going. Tuesday at midday, Duff, Harris, and Lambert talked. Lambert said it was decided then to move the oil spill group into the engineer's office. The phone company came in to put in additional phones Tuesday night. Lambert felt Duff and Harris saw that the oil spill group needed their own phones so that the station's salesmen could keep working. Wednesday, the oil spill group was all set up. Then the salesmen started saying they were not going to get off on the publicity. But Lambert felt it was evident the volunteers had a separate viable organization going. What they had at the station was two organizations working off each other and getting something from each other.

The station's union contract was up for negotiation in February and the first formal meetings were held February 16, 17 and 18. The business manager of IBEW Local 302 and the director of personnel for Metromedia were both present. The business manager of the Local said he felt they talked at those meetings about a lot of issues which were side issues for him, not bread-and-butter issues. They talked about the oil spill. They had gotten his daughter out for that one, saving the birds. The director of personnel said what he did in New York beforehand was to arrive at a dollar figure the corporation was willing to pay for the entire station. That was routine. Then, in the negotiations, it became a question of priorities at the particular station, a dental plan as opposed to a salary increase, for instance. But at those 1971 KSAN negotiations, they spent hours and hours philosophizing, talking about ecology and sexual freedom and FCC edicts regarding obscenity. He had never done that in a labor negotiation before. When he walked out of the last session in April, he felt both he and the business manager of the Local, whom he knew from 20 years ago when they both worked for IBEW in New York, had to admit they had not seen a negotiation like it before. It was something apart from them. The end result was the same as in other negotiations, but there was a challenge in it.

Representing the staff in the negotiations, along with the business manager from IBEW, was McQueen who was shop steward and a committee consisting of Street, McClay, and Bear. Duff and Campbell sat on the side of management along with the director of personnel for Metromedia. Campbell had been at the previous negotiations in 1969 when the rock staff settled on their first contract with IBEW and Metromedia. She felt herself an outsider on both occasions, more a spectator than a participant, and she felt there were differences between the two years. The first time, it seemed to her, the business manager of the Local was out of his depth. He did not know the format or what to do, so maybe he could not argue so well for the staff. By the second time, he could really fight for them. The second negotiations were more dramatic. The staff at one point acted like they wanted to call a strike. But they must have felt they could not get away with it. They had a lot to lose. Both the first and second time, it seemed to Campbell, the staff did not understand the negotiation process, that it was not all over right away but there was a procedure back and forth. The second time they wanted a two-year contract. The company wanted a five-year contract. It came out three. The director of personnel, Campbell felt, played it cool each time. He knew what he could give at the outset. The business manager of the Local would get red and sarcastic.

There was a suggestion of a strike just prior to the 1971 meetings. Duff wrote in his monthly report for January: "Preparations have commenced for the possibilities of a strike covering Engineering and Operations procedures." But even the business manager of the Local felt strikes were not the answer anymore. They took too much out of the ordinary man's pocket. He felt labor negotiations ran more on trust than they did on strikes and threats. After 26 years in the labor business and 15 years in San Francisco, he felt he knew the people he dealt with on all sides. He had to. And he felt it worked on the word. He knew when a guy meant it when he said something. When the director of personnel for Metromedia said no, he could tell when that meant a strike would not get them anywhere, or when it meant there was no use arguing further. He felt in this business you learned to roll with the punches, you learned to keep dirt in some corners.

This did not mean you encouraged people to ask for less than they could get. The staff at KSAN that year started out with a demand for $500 a week, a demand for instant arbitration, and for a two-year contract, none of which they got. The main item of controversy was a "no cause" firing clause which Duff and the director of personnel for Metromedia wanted written into the contract in place of the "unsuitability for program requirements" clause in the previous contract. The "no cause" firing clause meant pretty much what one of the station's engineers said, that a man could be fired for nothing on the drop of a dime. Not that this could not happen under the previous "unsuitability" clause, but with that one a man who was fired had legal recourse to arbitration to try to win back-pay. Young had taken his case to arbitration under the "unsuitability" clause. The business manager of the IBEW Local said he was for the "no cause" firing clause in the 1971 negotiations and he had changed in his thinking since 1969. In 1969, he had had doubts about the "unsuitability" clause, because even that seemed to put determination about firing too much in the hands of management, more so than with the previous contract. In the 1967 contract, which the rock staff inherited when they moved to KSAN, the firing clause was a "just cause" firing clause.

Between the 1969 and 1971 negotiations, the business manager of the Local felt it made a difference that they had the experience of Young. The arbitration in his case took nearly a year. It was finally settled in December 1970 with the arbitrator ruling that Young's discharge the previous December was within the provision, "unsuitability for program requirements" of the 1969 contract. The arbitrator said the basis for his decision was the judgment that Young had jeopardized the station in his treatment of advertisers and programming prior to the night in December when he suggested that listeners wire President Nixon with Hilliard's message threatening to kill him. The IBEW business manager said he understood the "unsuitability" clause in the 1969 contract was put in to enable the station management to get rid of talent when they no longer had appeal or when there was a format change. Putting a "no cause" firing clause in the 1971 contract would avoid a situation of having to take a case to arbitration to determine unsuitability as they had had to do with Young. Arbitration was costly and time-consuming and chances were the man lost. The arbitrator had to earn his keep and that came from both sides. The man would be better off just let go and given severance pay.

Duff said he felt it took a while to get the staff and the union to see that the "no cause" firing clause was necessary, because radio was not like assembly line industry. In radio you dealt with creative people and they were not interchangeable. Individuals mattered. The quality of their work mattered, and a union arbitrator was not in a position to discriminate that. Duff felt they wrote in a fairly generous severance pay clause in the 1971 contract and that the severance pay clause made the "no cause" firing clause just.

McQueen felt having the "no cause" in there meant the staff had no contract at all. He felt management put the elaborate severance pay scheme in to placate the staff. It bothered him that the staff just took it. They had a defeatist attitude toward the corporation. The prospect of going into arbitration to win back-pay in the end did not appeal them. McClay felt maybe their attitude was defeatist, but why stay at somewhere where they did not want you? The corporation would fire them if they wanted to anyway. Street remembered during the 1971 negotiations, the director of personnel for Metromedia, Duff, and Campbell said no repeatedly. The staff kept saying they should get more money because the station was bringing in more. Management said no, they were just breaking even.

The director of personnel took notes duirng the 1971 negotiations, more notes than he had taken in 1969. To begin with, he came out and told them the company could fire anyone it damn pleased. He came out and called a spade a spade. He told them the "no cause" firing clause would be in there instead of the previous "unsuitability" clause. He had in his notes that the staff wanted instant arbitration, a dental plan, more vacation, a Kaiser health plan changed to Blue Cross, and medical coverage for part-timers. But much of their discussion was not about that but about the political responsibility of the broadcaster. They wanted a say in the commercial content of the station. They wanted to be able to honestly say to their audience what they thought of the advertisers. Street was especially vocal on that. She was wearing a tee shirt which showed her breasts. The direct of personnel said he felt that made it hard for the IBEW business manager and himself to keep their minds on business.

In February, March, and April, while involved in the union negotiations, McQueen was also busy with the news. He wrote in his monthly report for March:

> February was the most hectic month in this department's history, and March wasn't much better. The month started with a bang when Weathermen set off a bomb in the U.S. Capitol. The next day we got "Weatherman Communique Number Eight" in the mail, postmarked Elizabeth, New Jersey. (AP and the New York Times also got copies.)
> The next day, the office of the Downtown Peace Coalition was broken into, and all their files stolen, including mailing lists and the files of who contributes and where they live. Several days later, KSAN learned that the police department had them, "found them down by the waterfront." They

were returned to the DPC, who inventoried them to make sure nothing was missing. They found, mixed in with the files, an envelope full of Christmas cards to the Police Intelligence Department from other police departments around the country, and even one from J. Edgar Hoover. We asked the Police Department if the Intelligence Bureau had gone through the files after they were found, and they denied it. We told them about the cards. Two hours later they called back belatedly to admit that Intelligence had indeed looked them over, and had somehow managed to mix in the cards. Intelligence Bureau?

Then the Black Panther Party split right down the middle. We had known for some time that that was going on, but couldn't document it. The split between the Oakland Panthers and Eldridge Cleaver in Algiers (plus the New York Branch), came out in the open. We immediately put a call through to Algiers and talked with Eldridge and Kathleen Cleaver, then called Oakland and talked with Huey Newton. We were the first to document what was happening, and report it fully with statements and conversations from both sides. Our story was picked up by the wires and many major metro papers around the country and in Europe (where it was covered more fully than in this country for the most part). We fed Washington, and then the calls began coming in. At last count we fed the story to more than thirty stations.

McQueen and the news department were also on the phone in March to Nicholas Johnson of the FCC concerning a notice on the broadcast of drug lyrics. The notice, issued March 5, called for the exercise of broadcaster responsibility in the airing of language tending to promote or glorify the use of illegal drugs such as marijuana, LSD, and speed. McQueen said in his March report:

> The FCC issued a controversial order regarding drugs and lyrics of records at mid-month, an order that is clearly a violation of every broadcaster's first amendment rights, not to mention the audience's and the performer's rights. We immediately got Nicholas Johnson, the only Commissioner to dissent, on the phone. What he told us was a stunner: That the order was prepared by and originated from the Pentagon. From a stack of records, that interview, and calls to the other Commissioners, we created a special, aired it a total of four times over the next few days. The response from the public was extraordinary. The switchboard was jammed for days; other stations asked for it, and aired it. It went out to other FM stations and a dozen or so stations around the country also aired it, with full credit of course.

Bonnie Simmons became the station's record librarian in January when Street became music director replacing McClay. Street said she got the music director job when she walked into Duff's office one day and told him she wanted him to fire McClay and give it to her. She knew Simmons and asked her to be her record librarian. The two of them then began to build the station's jazz file by getting old jazz records, and to expand the library as a whole by trading old promotion copies, three for one, for clean new ones they wanted. The station had been trading old promotion copies with record stores before, but Street felt she and Simmons increased the volume. She also felt having Simmons meant there was a stable record librarian at the station for the first time.

Simmons, who was 22, had worked at Music West before she became the KSAN record librarian. She handled the distribution of 50 lines and set up rack-jobbing for the record departments of White Front stores. She had contacts with record companies and knew album order numbers. She felt when she started at KSAN she already had a special kind of knowledge about the record business, and she felt she was a kind of media groupie. She was from Colorado. She had worked on some shows in Colorado in 1966 and had finished three years of college at the University of Colorado majoring in fine arts. In 1968, she ran away from home and came to San Francisco. She got the Music West job in April 1969. In May, she married Bob Simmons who worked part-time as a disc jockey at KSAN. He heard about the record librarian job when it opened up and told her.

Simmons felt when she started the station had a bad reputation with a lot of record companies in town. The FM rock stations had a bad reputation generally and a promotion man might not know KSAN was different. And there had been incidents of people ripping off records. One of the things she did in her first few months was to go to the distributors and try to make a good impression and educate them about how they worked at KSAN, that they did not list a top-10 each week like some of the other stations, for instance.

Harris hired a new black salesman in January and Donahue came back from Europe from his Warner Brothers movie-making trip and took up his old Saturday night spot. The fate of the movie he had gone to make was unsettled, but while in England, on December 5, Hamilton had given birth to a boy, her first child, his fifth. The Warner Brothers movie trip they had left on in August had ended them up in England in September and they had stayed on. The movie was shot by a French filmmaker and was about a caravan of about 150 people Donahue had assembled in July, 150 San Francisco freaks and friends who took a three-week trip across the country. Among them, in addition to Donahue and his family, were Wavy Gravy and a group of Hog Farmers, Melvin, Laughlin,

Towle, and the Congress of Wonders.

The caravan had travelled across the country in an assortment of trucks, buses, and vans, camped out in tie-dyed tepees, and thrown free concerts along the way. The concerts featured Warner Brothers artists who were flown in on scheduled occasions to perform. The lead van had painted on it an inscription Donahue was said to have chosen as the unifying theme for the caravan: "We Have Come For Your Daughters." A writer from the *Village Voice* and one for the *Underground Press Syndicate* reported on the caravan and criticized it for being a rip-off of the counterculture and a sell-out to Warner Brothers. Warner Brothers had also produced the movie, "Woodstock." A writer from the *Stone* went along at Donahue's request to record a history of the trip. The account he finally put together ran a statement by Donahue toward the end:

> If there's one thing I find personally satisfying, it was the casting job. I think we came up with a true cross-section of the cream of the freak community. Because I think I know as many freaks from the Bay Area as anybody in the world. And they were some of the choicest, man. People whom I consider really committed to a life-style. In the end I think it was largely because of those people that the caravan stayed flexible enough, loose enough, so that no one could really take it over....
>
> Everybody has to do their role. I don't pretend to be a hippie, because I don't think I am. I think I'm a different variety of freak. We are each different varieties of freaks. If I had to come up with a definition I'd say I belong to a group pf people who are not of the same age as the freaks but we sympathize with their aims. We become the accomodators of their pursuits. With me it's been accommodating musicians. Or this past summer, accommodating the caravan.
>
> Frankly, that's one reason why I avoided Francois' cameras once we got rolling on the trip. I could see that he wanted to make me into a character, a narrator, an explainer, whatever, and I just couldn't get behind it. Because that wasn't my role. I wanted him to concentrate on the reality of the people on the caravan.
>
> It gets back to the presentation of the truth. I think one reality the caravan demonstrated was that the revolution has already passed the turning point. Not necessarily because of what happened on the trip but because of the people on it. Many of them were a direct outgrowth of a scene which developed years ago, which evolved into a life-style, a functional philsophy which has already survived the major test of its viability. And they represented a larger group of people who have adopted that life style....
>
> I'd like to be able to say the movie that will come out of the trip will be a good one, but it's too early to tell.... In a way I'm really not that concerned about the movie. That may sound like I'm contradicting myself, but

it goes back to the role each of us chooses to play. Because while I'm intrigued with the idea of properly recording history, I wasn't the one who was doing that recording on the trip. My principal role was to bring together a group of people which to a great extent represented a larger culture — with both its strengths and weaknesses. You could say the caravan had an artificial beginning because Warner Brothers provided us with the means to create it, but that never really bothered me. Because I knew that as soon as we pulled out of San Francisco, the caravan would become very real. The real events would occur. And it was up to Francois to document those events. We all would have liked him to have approached some things differently, but its out of our hands now. We'll just have to hope that his movie will turn out to be our movie.

Late in June, about a month before the caravan was scheduled to leave, Donahue had announced on his Saturday night show on KSAN that listeners who wanted to join it should write him care of the station stating why they wanted to go, what they could contribute, and enclosing a photograph. The writer from the *Stone* said the station received 300 letters from listeners within eight days.

A new rate card was put into effect on January 1 and another on May 1. Duff wrote in his monthly report for February: "Our requirements for restocking our 'goodie closet' are critical. I have spoken with Gerry several times (including a request for some radios needed for merchandising in advance of the consummation of the trade deal). No results yet." Also he said he felt it was time to consider another special college campus Pulse. Metro Radio Sales was using the old one frequently. The station ran a live remote B.B. King special from the Showcase Club in Oakland in February, a special on the anniversary of the death of Malcolm X, and one on the radicalization of Timothy Leary by Baba Ram Dass. Boucher said in his monthly report the Rollings Stones' new movie "Gimme Shelter" came to town. Ponek was in it doing the initial narration. There were excerpts from KSAN's Altamont broadcast, a shot of McClay interviewing Jagger, and a KSAN credit at the end.

Monday, March 8 was International Women's Day and Boucher said they put an all-girl staff on the air for the day. Voco left in March to go to work for the ABC station, KSFX, previously KGO-FM. He had been doing weekends on KSAN. Boucher said Voco was the fifth person to be lured away from KSAN by KSFX and that KSFX was spending a lot of money for their air staff. The others they had taken were Pigg, a promotion director who went as a producer, the husband of Simmons who had been a part-time disc jockey, and one of the salesmen. The January-February ARB came out in March showing the station down slightly.

Boucher said in his report the January-February ARB had been taken during the oil spill. The cleanup seemed to have affected the ratings negatively but he felt it had made for much good karma and would result in larger future audiences. The Joe Cocker movie, "Mad Dogs and Englishmen," came to town in March with a scene showing Donahue interviewing Cocker and Leon Russell in the KSAN record library. Boucher said, alas, KSAN received no credits in the movie.

On April 3, Duff replied to a letter he had received from a listener complaining about the station's conduct after the oil spill. The listener had written on March 30:

> KSAN 'People'
> Wow, are you shit!!
>   Since the Standard Oil collision, you've been putting down all this good stuff about picketing Standard Oil and what bastards they are for cutting off funds for the Bird Center — then tonight I hear "Does F-310 Really Work?" on your station. Bet you got a lot of money for running that commercial, huh?
>   Good for you!!
>   Your station spouts really together rhetoric — so it's quite a bummer to discover that's all it is.
>   Don't ease your conscience with the anti-pollution nature of the F-310 ad — The Federal Trade Commission has a suit against Standard Oil claiming that the anti-pollution cliams for F-310 are bullshit. How does it make you feel to know even a federal agency has more balls than you?
>   Who owns Metromedia?
>   Who owns you?
>   Tom Knutson

Duff replied:

> Tom,
> We get the same money from Standard as from the most righteous sponsor going. Keeps us alive and on the air. So we can do what we want to do. Metromedia is publicly held. I am privately held, balls and all.
>   Kindly,
>   Willis Duff

Duff may have included a copy of the letter from Knutson along with several other letters of complaint from listerners about commercials he sent to Duncan soon after with the note: "I get dozens of these a week. Willis." Duncan became president of the Radio Division as a whole in April and shortly thereafter appointed Duff vice-president in charge of the West Coast stations: the two in San Francisco, the two in Los Angeles, and two in Cleveland. Duncan said he felt there were programming problems with these stations and Duff was good at that.

Duff said when Duncan became president of the Radio Division, Duncan asked him at first to be his programming assistant for all of radio and program director for WNEW in New York. But he told Duncan no. A few days later Duncan called back and offered him the West Coast regional vice-president position which he accepted. He still kept his managership at KSAN. It was Metromedia corporate policy to economize on personnel by having their administrators work also at local levels. Duff said he felt the regional vice-president appointment shifted his interest away from KSAN, but that was all right because about that time he was becoming bored with just the station. The same things kept happening over and over. It was repetitive. The staff had the same complaints they always had about commercials. In meetings he could almost predict how they would act and who would take what roles. He had been at the station two years. He felt the regional vice-president job would give him a chance for a change, and a chance to expand his experience, and he felt he could delegate authority to others at the station when he was not there. He could leave Harris in charge.

Chapter 12

KSAN
(June-October 1971)

The sense during these months is of having resolved the underground/establishment dilemma in a way more favorable to the establishment than had been apparent previously. This is to be seen in new hirings, a bombing incident, and response to criticism from the underground press. What is prized especially by this time is a competence in dealing with the world which, while serving the station, is not specific to it. The nature of this competence is indicated in discussions of professionalism.

Harris was off on vacation in June. Gardner filled in for him as sales manager and wrote up the monthly sales report. In it he said the April-May ARB had come out and was a down book for the station. But neither KMPX nor KSFX had made significant gains. He felt listeners had temporarily switched to baseball on KSFO (Giants) and KEST (A's), both of whom showed considerable gains during ball-game times, and this led him to ask why KSAN could not do some kind of sports.

The biggest programming event in June was the closing of the Fillmore West, Bill Graham's rock concert and dance hall on Market Street. From Tuesday, June 30 through Sunday, July 4, KSAN broadcast live the final week of concerts, five hours each night, from 9 p.m.-2 a.m. The first two nights the broadcasts were in quadraphonic in conjunction with KSFX. There were some technical difficulties. Duff said in his June report, these were yet unexplained but the encoder was taken out of the circuit and for the rest of the week, KSAN and KSFX broadcast the concerts separately. KZAP in Sacramento also shared the feed. On Monday night, the first night of the Fillmore's closing week, there was no concert scheduled and Graham came into the KSAN studio, played records, and talked with Street on her show. Tapes of the concerts during the rest of the week were mailed down for rebroadcast on KMET in Los Angeles the day after they ran on KSAN.

Many reasons were given for Graham's closing of the Fillmore West at the end of the week of June 28 and his closing of the Fillmore East the following week in New York. McQueen said in his news report someone had dropped LSD into punch which was handed out at Winterland during a concert and the resulting uproar was the last straw which led to Graham's quitting the business. Graham said it got to a point where he felt he had become a victim of success, where running and financing the Fillmores were dictating his lifestyle. Gardner said in his sales report the week of Fillmore closing concerts broadcast on KSAN was sold for a total of $8,800 to six sponsors: Music Odyssey — a record store, Columbia Records Sales, Coca Cola, Topps and Trowsers, Warner Brothers, and Nunn Bush for their Brass Boot retail stores. Over 50 promotion announcements were run on the air carrying the sponsors' IDs. Duff said in his report the $8,800 the station got from selling the closing gave a nice boost to the June sales gross, but the concert was very expensive to produce. He also said that the closing was an end-of-an-era phenomenon. Graham had held his first rock concert at the San Francisco Fillmore in December 1965.

Harris left the station in June to become sales manager at KMET. He felt he had been coasting with the sales success of the station for almost a year and in some ways it was getting boring. When Duff offered him the sales manager's job, Harris said he felt he would like to see how he would do in another market and he thought it would be trying a corporate career as opposed to the more independent things he had done before. He made one last appointment to the KSAN sales staff before he left, a salesman who had a different kind of background than that of any they already had. Dymond said in a press release his most recent positions prior to KSAN were as publisher of *Avant Garde* and *Moneysworth* magazines in New York. He was previously managing editor of *VIP* magazine, associate editor of *Playboy*, associate editor of *Eros*, and editor of *Chess World* and *Chess Life*, and from what Dymond personally knew, he was once a school buddy of Harris.

Dymond in June was making plans for two KSAN promotion campaigns to run in the fall. One was a bus comix promotion and the other a billboard competition. She said in her monthly report she had spoken with Victor Moscoso about designing a set of bus comix for the station. Moscoso had designed the first KMPX poster in 1967. He had been a rock concert poster artist in the mid and late sixties and more recently was drawing *Zap Comix*. He had agreed to begin work August 1 on a set of KSAN cartoon bus panels to go up inside San Francisco Municipal Railway buses in mid-September. Dymond said she had made arrangements with Metro Transit, a division of Metromedia which handled the selling of advertising space in the buses, and the station would pay

Moscoso $1,000 for reproduction rights. In 1967, North Beach Productions had paid him $150.

For the billboard competition, the second fall promotion, Dymond was proposing a contest open to anyone who submitted a design for a billboard advertising KSAN. Notices giving specifications and soliciting entries would go out in September. Entries received would be judged by a jury in October and handpainted copies of the winning design would go up on three queen-sized Foster and Kleiser billboards in the Bay Area in November. Foster and Kleiser, too, was owned by Metromedia. Runner-up entries would be exhibited in a local art gallery.

The billboard design competition was opened to the public on August 25. The on-air promotion was a 60-second spot of audio collage by the Gestalt Fool and live copy with billboard specs. It was scheduled to run once each show for the length of the competition. The prize offered for the winning entry was $1,000 cash plus reproduction on three billboards. As of the end of August, 3,800 entry forms were out.

On Sunday, September 5, while the billboard competition promotion was in progress, a bomb blast reportedly went off at 7:07 p.m. at the Foster and Kleiser headquarters in Oakland, shattering a concrete floor, a receptionist's desk, a telephone switchboard, and stained-glass windows. Foster and Kleiser subsequently claimed damage of about $40,000. On Monday, September 6, the day after the blast, KSAN received a letter from a Revolutionary Army which took credit for the bombing. The *Chronicle* ran an article on September 8 reporting the event and quoting from the letter:

> The "Revolutionary Army" took credit yesterday for the Sunday bombing of the offices of the Foster and Kleiser billboard firm at 1601 Maritime Street in Oakland.
>
> In a letter to radio station KSAN, the group said that "billboards in Babylon are an offensive manifestation of pigthink. Their fascist distortion of our people's reality can no longer be tolerated."
>
> Babylon is a term for the United States coined by Black Panther Eldridge Cleaver. Both KSAN and Foster and Kleiser are owned by Metromedia, Inc.
>
> The blast, which went off at 7:07 p.m., blew a five-foot-wide hole in a four-inch-thick concrete reinforced floor. Stained glass windows were shattered, a telephone switchboard was blown out and a receptionist's desk smashed.
>
> The letter also suggested that KSAN cancel a billboard contest it is currently running — a $1000 prize for the best design for a billboard advertisement for the radio station — and instead "donate the money to a revolutionary need."

"Billboard companies are ordered to withdraw from the People's Community of Alameda county," the letter declared. "Foster and Kleiser, whose office is now a crater in the midst of the Oakland Army Terminal, has just received the first warning."

A Foster and Kleiser spokesman had no comment on the letter, but estimated the damage from the blast at about $40,000. The letter has been turned over to the FBI.

The *Associated Press* carried a similar story and versions of it appeared in local papers. The *Tribe* of September 10 reprinted in full the text of the letter from the Revolutionary Army, called the letter a communiqué, and said it had arrived at the station slipped in with the regular Monday morning mail.

McClay said he thought the KSAN staff felt great that Foster and Kleiser was bombed but they knew it was not good for the station. Dymond said Metromedia in New York wanted the station to cancel the billboard competition after the bombing, they were angry, but Duff gathered up an armful of posters and went over to Foster and Kleiser in Oakland and talked with them. From what she knew, he told them he thought the station ought to have local autonomy. He told them he thought the bombing was a one-time threat and they should save face and not back down from the bombers. He did make one concession though, no billboard in Oakland.

Duff said the initial reaction of Foster and Kleiser was that it was all KSAN's fault. They were hysterical and they had no sense of why anyone should be angry with them. They had no sense of ecological offense. They felt the whole of the source was the radio station. He had to deal with the logic of their argument. Duncan said Foster and Kleiser assumed in Los Angeles that the radio station had influenced the bombing. But that was not so. Foster and Kleiser could have been bombed by mistake instead of the armory across the street. Duncan said he felt there was no cause-effect relationship between the station and the bombing. The explanation was more complicated. When Foster and Kleiser reported to the president of Metromedia, they said it was the station that did it. Duncan said he told them that was emotional, the armory was up the street. When Duff called, he told Duff to go over there and placate Foster and Kleiser. The whole incident was an example of overreacting.

Boucher resigned as program director in July. He said he decided it was too much work, too much hassling. He knew a program director job was a prime job for most people. They could build up their reputations as program director. They would seem in their monthly reports glowing whatever their stations did. But after a point he knew he did not want the bureaucratic work, he did not want to make it in the Metromedia hierarchy.

Bear said Boucher resigned after Street called him one night and balled him out. Bear said he himself had been talking with Boucher about the state Boucher was in, how it was down and how it was bringing them all down. Street said she called Boucher and berated him. He asked her, "Do you want me to quit?" She said yes, and the next day he quit. She was the last straw.

Boucher said in July he just quit. Maybe the last straw was preparing for the week of Fillmore closing broadcasts. It was a lot of work.

The man who replaced Boucher as program director in August was Thom O'Hair. O'Hair was, by his own description, a cause-effect man. Boucher, he felt, had been too much a hippie and too namby-pamby for the job. It was Bear's opinion that O'Hair was good for the station as program director at the time he came. He was not ideal and not a real heavy, like Pigg might have been, but he did have balls.

McClay said he thought Boucher in doing nothing had been an ideal program director. He did not interfere. But Boucher did not feel he did nothing. He felt duirng his time as program director he kept hassles away from people, he got live music going, he hired and fired, he did reports. The staff, while he was program director, became less violent than they had been under Ponek. But he could see when O'Hair came that O'Hair had a different philosophy.

Ponek said when O'Hair came, he seemed shell-shocked and the staff was out to make it hard for him. But O'Hair had balls and Ponek respected that, although he did not trust him. Ponek felt it was O'Hair's style not to appear indecisive. He would tell people, "Fuck no, man, that's stupid," when he thought so. He seemed very in touch with what he was going to do. He was a foreman type, an executioner. He could bullshit well, and bully, and imagine all sorts of scams. Duff liked him. He was a kind of fantasy hero for Duff.

Laughlin felt they had no program director at the station before O'Hair. But it was not clear they had one afterward. Like Boucher, O'Hair's official title was operations manager and he took on an administrative overseer role. That was not what Bear had in mind when he thought of the ideal program director. The ideal was the kind he would be if he had the chance, someone who stimulated programming ideas among the staff and set an example with his show.

O'Hair was 29. He was married and he had a 5-year-old son. He had been born in Chicago and had worked in radio since he was 19. He was closest as he grew up to uncles who were fighter pilots and he liked having a model airplane on his desk to remind himself to be prepared. He had been working at an FM rock station in Portland, Oregon just before he came to KSAN. He had taught for a while at Chico State College and had started work there on a Ph.D.

O'Hair said he had worked weekends at KSAN the winter before he came on as program director and had then gone back to school. Boucher called and asked him, but he was McClay's friend previously. He met Duff that winter. Then in July after Boucher resigned, Duff called him in Oregon and asked if he wanted to stay there all his life. On July 21, Duff flew him down to San Francisco to talk and he agreed to take the program director's job. When he started at the station in August, he thought there would be more order, more rhyme and reason in the way things worked than there turned out to be. It seemed to him what he found was that what was going on was completely accidental and day-to-day, and he got a changed impression of Duff. He realized Duff had no balls, that Duff let people walk all over him and that he could be threatened. He also realized that the station was a fantasy, a fantasy of hip. It was not reality. You could act like it was but that did not work. What went on the air was completely different from what went on behind.

When Boucher resigned and O'Hair replaced, Boucher expected to go back to the job he had before producing the station's commercials. In the agreement he had made with Duff when he took the program director's job, he said he was assured if he did not like the job, he could go back to his old one in production. Now, a year later, for him to go back meant the man who had taken his place in production, Roland Jacopetti, would have to leave. Jacopetti said he was shocked when Duff told him he was going to have to fire him for Boucher to have his job back. He got four-weeks notice, extended to five weeks, and he had three kids. He did not remember having been told that Boucher had rights to have his job back if he could not cut it as program director.

After some complication, Jacopetti left KSAN and took a job at KSFX. This particular job was one Katie Johnson had just quit. Johnson said she quit because she felt the job was horrible. She had taken it because it was in radio, it was working with equipment and it paid $209 a week. She had kept it from February to September and when ABC sent down instructions to take out dirty words, she took them out. Once for a fuck, or maybe it was a shit, she substituted a sound effect that sounded just like it. That was very different from what she had done in her early days of producing poetry-sound collages at KPFA. Johnson had been Donahue's chick engineer at KMPX, had quit KMPX a few days after he did, but had come back for the KMPX strike and been a member of the strikers' negotiating team. The summer after the strike was over, she had gone back to KPFA and turned down a few jobs Donahue offered her at KSAN. She still had bad feelings toward him, she felt she had been used by him during the strike. At one point he offered her a Sunday night shift on KSAN. She was doing a Sunday night shift on KPFA at the time and she felt she would be more limited on KSAN so she turned him down. Then she took off for Europe.

She came back in May 1969, soon after Donahue's wedding, but she felt apprehensive about going back to KSAN because of the possible pressure to take dope. It seemed to her Voco and Street were weird then. During Christmas of 1970, she worked for Voco and Street for a week or so filling in for them and later filled in for them occasionally. Then in February 1971, she took the commercial production job at KSFX. Street still asked her sometimes to fill in for her on KSAN, but she was too tired. Then they got out of the habit of asking her. At the time she quit KSFX, she was tired of commercial production and she did not want to get messed up with management politics and the bad morale at the station. She heard she was replaced by Jacopetti after she left.

Street remembered Johnson from KMPX fondly in some ways but cautiously. She did not give out Johnson's phone number without thinking Johnson might not want to be disturbed. In July, shortly before Boucher resigned as program director, Street resigned as music director at KSAN, keeping only her air shift. Bobby Cole took her place as music director. One story was Cole got the job by playing with Street's tits. Street said what happened was she finally got burnt out. She was working days as music director and then at night doing her show, it seemed like she was working all the time, and Cole was the only one there who had not had the music director job before. She knew the music director should get paid more, so she got it raised from $25 to $75 a week when he took over.

Cole had started as a weekend disc jockey at the station in the fall of 1970. He had been doing a shift on KMPX and had quit two days before the KMPX lock-in in September. He was a friend of Gossett's at KMPX and, like Gossett, was raised in San Francisco and about the same age, 23. He had gone to the College of Marin and San Francisco State College, taken radio and broadcasting courses, and in 1968 dropped out. When Donahue was on KMPX, Cole had called him once and asked how he could get on. Donahue told him, "Does your father have a lot of money?"

Simmons remained record librarian after Street resigned. She knew with Cole some things would be different. He liked monster music, cock rock, not rhythm and blues like Street. But her job had changed by this time. In January when she started, the record librarian job included handling public service announcements and answering the phones at lunch. She bitched at Duff that they should hire someone else to do public service. Finally in August he hired someone for public service. She unloaded the phones in April. At first she felt she learned a lot by answering the phones. People who could not function at all called the station. But it was a burden. When they hired Campbell's assistant in April, the assistant took over the phones three days a week.

When Cole replaced Street in July, Simmons was calling in airplay reports to *Record World* and *Billboard*, but to *Billboard* sporadically because they toyed with the list. She was still secretary to the program director in addition to being music librarian, which she said meant she had to type monthly reports. She learned how to type doing that. Boucher was spaced out a lot and she thought maybe she started doing more because of what he was not doing. By the time Cole became music director, she felt she had gotten her system set up. She had agreed with Street's selecting Cole for the job. There was nobody else to do it. In her system, Simmons did most of the ordering. She had relations with the record companies. Cole as music director had more time to listen to records than she did. But actually they shared a lot of it. There was not such a great distinction between their two jobs. In many radio stations she knew there was. In many stations, the record librarian just collected the records. But in her case the job was expanded, she had expanded it, and now what they had was two music advisors.

Sanders thought it was sometime in June or July that Duff split with his wife. It was just before O'Hair came. Duff laid low for a while, then he did wild things, fucked around and drank, and things changed at the station. Duff said the real split came later. Rob Skinner said Duff called him the last week in July and said he had been thinking of hiring him away from K101 where he was sales manager. He said he had been thinking about it for a long time but he had not had the money. Skinner recalled he had met Duff in meetings in Los Angeles. He came over and talked with Duff that week and the next week, the first week in August, he started as sales manager at KSAN.

Skinner bought himself an electronic desk calculator his first week. Harris had not had one. Skinner felt Duff needed someone to replace Harris who was not a hippie or a pseudo-hippie. He needed someone with a professional attitude. Skinner felt he had a professional attitude and that he had a track record from his three years as sales manager at K101. When he came to KSAN, he found that the salesmen were not professional, they had a lax attitude. He had to educate them about using numbers and computing ratings and costs. In late August and early September, he held seminars with them with people from the rating services. He started taking them around to agencies and introducing them to the Ad Club and Milline.

He encouraged them to get to know sales managers of other stations. He told them he felt all relations between competitors were not competitive. They were all getting the same crap. He did some things to instill professionalism by implication. He tried to set an example in how he dressed rather then telling them what to wear. He emphasized promptness. He wanted the salesmen to get to work by 9 a.m. each morning, before they went out on business, instead of wandering in later after they had made a call or two. He wanted them to be places on time, especially when they made appearances at agencies. Probably 40%-50% of the accounts they had were agency accounts and they had not been getting there on time. He told the salesmen if they wanted to get jobs with other stations they were going to have to shape up. He talked about professional ethics and fairness with them and doing what was equitable. He felt things were top-heavy at the station when he came. The salesmen who had been there the longest had the better account lists and made more money. It was not equitable. He divided up the lists so they would be more even. Harris, from what he knew, had the opposite attitude, that the longest there deserved to make more.

Skinner felt it was frustrating working at the station, especially at first. The attitude the staff seemed to have was that things would get done. He felt that was a false attitude. If you wanted something done you did it yourself. He felt, like O'Hair, an antagonism to the hippie ideal he assumed they had. Maybe it was appropriate before, but now it was bothersome. It got in the way of carrying on business day-to-day and it clouded the sense of what they should be doing. O'Hair felt when he first came he sold himself to Duff with his analysis of what was wrong with the station and that included a feeling he had that where they came from, KMPX, the beginning, was a beautific world. It could not last.

Gardner recalled when Skinner came he started right off educating the salesmen to do agency business. Harris had not done that. Harris had been a street-scrambler type. Skinner started educating them about using ratings so they would understand the methods, and he forced them into the agencies. Soon after he came he gave them assignments of two agency lunches a week and told them to get to know everybody there. He would take away accounts from them if they were not doing well. About the numbers, Gardner felt the education Skinner gave and the seminars the salesmen had with the people from ARB and Pulse probably gave them self-confidence more than technique. But they were doing better in ARB by then, so maybe it would have come anyway.

Dunlop felt he had conflicts with Skinner but that Skinner was a good type for the job. Dunlop said Duff asked him to be sales manager when Harris was leaving but he did not want it. He told Duff he had sat behind Duff's desk and Harris' desk and he did not want the sales manager's job. He wanted a national sales manager's job. By that time, the summer of 1971, Dunlop felt he had really come up. He was making the most money of all the station's salesmen. He was no longer the youngest, he was second youngest. He felt like Horatio Alger's Matchbox Kid. He had read the story and he could really identify with it. The Matchbox Kid, that was him.

Nemerovski felt there was a gap between when Harris left and Skinner came, about a month, and then when Skinner came things changed. Before Skinner, the salesmen had always been judged by their billings. With Skinner, they were judged on their professionalism as well, how they looked and acted and dealt with expectations of agencies. Nemerovski felt Harris had been an aggressive conceptual thinker, he could think on his feet, he believed in the radio station and what it represented. Skinner was more a manager. He was interested in contacting agencies. He was a numbers man. He could tell you about cumes and quarter hour shares which Harris never knew about. He had a different karma than Harris. Skinner was 28.

Campbell was 32. She felt it was a relief to have Skinner replace Harris in August. Skinner was a businessman, whereas Harris had been self-aggrandising. All he wanted to do was make himself look good. Campbell intuitively understood Skinner. Like her he was a Virgo, compulsive and particular about his job, and he did not imagine himself a hippie. He saw himself as different from most of the other people at the station and he felt it was important to keep his privacy. Campbell did not understand Harris as well. She and Harris had always had fights. She felt Harris was an insecure man, that he was friends with Duff, and that he tried to father figure Dunlop. He let Dunlop get away with a lot. He let him sell accounts they could not collect on and he let him develop sloppy habits. As a result, Dunlop was selfish, inconsiderate, and egotistical, and Skinner had to have it out with him.

Campbell believed both Harris and Dunlop had the intelligence to do better and it angered her that they did not. Harris did a lot of trading out and he would get accounts they could not collect on. The collections problem dogged Campbell. It was always there. She said sometimes she tried to do something about it, but she did not like to do it so she did not do it. Harris did not do it and the salesmen did not do it. It was one of those frustrating things. She would hang up the phone after making a call about collections and start swearing. She had never said "fuck" before she came to this radio station. She did not raise her voice at home. She was the quiet one among her friends. But the environment around the station made her feel she had to be aggressive. She felt it dated from the early Duff period when they had those encounter-group staff meetings where no one listened to anyone. She started becoming this other personality then. At one point she wondered if what was happening to her personality was good. She was one person at home and another at the station. Actually, she was not two people. It was just another part of her that came out here. She had to be like that to survive.

Skinner had once worked as a top-40 disc jockey. He had not pursued it, he said, because he felt the personality thing was short-lived. There was more security and more future in sales than in being a jock. As a disc jockey you had a limited income and a limited future unless you were exceptional. Skinner did not think he was exceptional in the way a successful disc jockey was, but he did feel he had a personality and that it was one people liked, and he liked dealing with people. He did not like corporate bureaucratic paperwork. Ideally, if he could have his way, he would be in business on his own. He had made plans with a friend while he was at K101 to buy a station in Sacramento if and when it became available. In that way, too, he was like O'Hair. O'Hair had plans in the back of his mind for sometime in the future making it on his own by buying up partnership interests in a number of radio stations around the country.

O'Hair, probably more than Skinner, felt he was on the line from the start when he came to KSAN. Beginning in August, for about six months, he felt an evaluation of him was going on at the station and it was going on behind closed doors. He had not expected that. When he talked with Duff in July and said he would come, he had not expected his first main concern would be staying alive and surviving. Street said O'Hair had to prove himself to the staff. He had to prove he was for them against the outside. They put him on trial like they did everyone. They put Skinner on trial too.

Nisker came back in August, after having been in India for about a year and despite all the hard feelings he had when he quit. He and Duff still liked each other and KSAN was a place he felt he could go. He started doing news features again, twice a week, in August. He was not told what he could not do. He felt he did not need to be told. His attitude was not as political as it had been when he left. He had come through some changes with KSAN. If he had to date them, after he got to a certain level of technical proficiency, first was his show business period, in 1968-1969, when he thought the station was doing a good thing and his collages were like theater pieces. Then was his political period, in late 1969-1970, when he became radical to the point of producing shows which were incendiary, one of which ended up in his quitting. Then he split and went to India. Now he was back and not very clear about what he should do but thinking he might get political again.

Bill Kibler also came back in August. Kibler, an engineer, had been away since April 1969. He was 25. When he first came to work at the station, he had just finished at the College of Marin where he went after getting out of the Navy. He had learned engineering in the Navy and considered himself an electronics engineer who got into broadcasting, whereas usually radio engineers were broadcast people initially. They were broadcast technicians who then got into engineering. Kibler said when he first came to KSAN, Metromedia had an overall chief engineer for both KSAN and KNEW. This overall engineer had redesigned the construction of KSAN after Metromedia bought it from the independent owner in 1966. He redesigned it with a minimal investment because the Metromedia policy generally was to buy up stations, slightly upgrade them, and turn them over.

Kibler felt many of the engineering problems they had at the station dated to original design defects the overall chief engineer had put in in 1966. It was not that the overall engineer was incompetent, but he was the type who could not be told he was wrong, it had to be done his way. When the rock group came on, he set up politics of maintenance and established as an attitude of the engineering department that KSAN would not last more than six months to a year, so nothing major should be done in the way of engineering improvements.

The station had lasted three years by the time Kibler came back and the overall chief engineer had gone. Kibler replaced a previous engineer who had been there when the format was classical and in the beginning with the rock group, but had not gotten along with the rock group because he hated their music. Kibler felt one way he differed from the man he replaced was by starting simple routine maintenance and fixing things regularly and at night. The previous chief engineer had fixed things on his overtime and done maintenance in an ad hoc way. By coming in at night, Kibler felt he could work without trouble from the staff. A problem at the station was there was a clique and unless you were in with them, you had trouble. By working at night, he got to know Street and the night people and it was easier to get things done.

Another thing Kibler said he did was to start correcting some of the construction mistakes the overall chief engineer had made when Metromedia redesigned the station in 1966. The first major correction was to rebuild studio two, the main broadcast booth, to meet FCC requirements and make it more usable for the staff. It was an FCC regulation that an operator had to be able to see all controls from his normal operating position. The control board was supposed to face racks of remote equipment. In the KSAN studio it did not. Kibler said this was brought to his attention at a conference of local radio engineers he went to. Some of the other engineers brought it up as a thing that had to be changed at their stations. He came back and told Duff and Duff was willing for him to change it and rebuild the KSAN studio starting in the fall.

On October 18, the station again received a bomb letter, this time from a Sam Melville Squadron of the Revolutionary Army. The Sam Melville Squadron claimed responsibility for the bombing of an Iranian consulate in San Francisco on October 14. Their letter was sent to the station special delivery and dated October 15. The *Examiner* of October 18 reported that KSAN had received the letter and that it had apparently been printed with a toy printing press. The *AP* of October 18 ran an item on it. The *Barb* of October 22 ran a copy of the letter as printed along with an article about the Revolutionary Army and the Iranian consulate. The *Tribe* of October 22 ran a letter of their own, addressed to KSAN, which was critical of McQueen for turning over revolutionary letters and communiqués to the police and the Feds, even when they were xeroxed copies. It ended with the statement that "KSAN should realize they aren't just making news — they are making revolution."

On the subject of KSAN's role in the revolution, the *Barb* the following week ran a front page headline, "McQueen Comes Clean, Does KSAN Equal KFBI?" and an article which focused on McQueen's thinking about what the station should do with bomb letters. McQueen said he thought the *Barb* article was good except for the headline. It began:

> Radio KSAN makes available to police or FBI, "if they ask," letters and other material sent to the station by political bombers, news director Dave McQueen told Barb Friday.
>
> To not do so, McQueen argued, would be to support bombings, and "what ultimately has to be said is that we do not, ultimately, necessarily support bombings."
>
> To its credit, the station does not on its own initiative send material to police agencies.

But, "if they ask," letters and other communications received claiming credit for bombings or other political violence are routinely handed over to the FBI or SFPD.

Under this policy KSAN has delivered "several" bombing notes to authorities over the past few months.

"These groups, goddamnit, they know these things are going to end up with the police," McQueen said. He mentioned that often the same letter is sent to the Chronicle or AP, which clearly could be expected to do the same thing that KSAN does with them.

"To do otherwise would be a flat out statement that we approve of (bombings), and we're just not in a position to do that." McQueen continued. "We're a federally licensed radio station and we can have our license removed by the FCC."

The *Barb* article went on to say that the news director for KPFA said KPFA would never make anything available to the police except under court order, and it was his understanding this could not be prima facie grounds for license revocation. He cited CBS' refusal to give out tapes of "The Selling of the Pentagon" to the Senate investigating committee as a precedent for withholding material. McQueen, however, felt the CBS precedent did not apply, because the material KSAN surrendered was material that had already been read on the air, was public, and hence, could be subpoenaed. It seemed clear to him that KSAN was legally required to provide transcripts and possibly even a tape of the news broadcast in which the material appeared. But it was not clear the station had to provide the piece of paper itself, which was possibly fingerprinted or otherwise traceable.

The *Barb* reporter who interviewed McQueen for the October 29 article asked him if the station felt any qualms why they could not simply destroy the originals. McQueen's response was to say he had been wrassling with that question from the beginning and the more he thought the more he was convinced they could not destroy them. To do so would be an overt act of support for the persons who did the bombings. The *Barb* reporter then asked if handing over the material would not been act of overt support for the FBI or police. McQueen said no. He went on to say if a bomber showed up at the station, personally, to give his side of the story, the station would not call the police. If the police came, he personally would not give a description of the bomber or a tape of the conversation because of the possibility of voice prints, but he would furnish a transcript of any conversation aired.

On Saturday morning, October 30, McQueen talked on the air on KSAN with reporters from the *Barb*, the *Tribe*, and the *Good Times* about the station's policies on turning over bomb letters to the police and the FBI. He told the reporters he had changed his mind on the subject and that in the future he would not turn such letters over. They were on the air during Laughlin's talk show, which had been moved to Saturday mornings from Sunday nights, and Laughlin, as Travus T. Hipp, took part in the discussion. The following week when the *Barb* came out, there was an article by the *Barb* reporter present:

> The news department at radio station KSAN won't be handing over to police and FBI bombing letters received by the station "until we get a clear and concrete ruling as to exactly what our responsibilities are, from Metromedia (which owns the station), and from the FCC, and from the courts if necessary," KSAN news director Dave McQueen promised on the air Saturday morning.
>
> "The next letter that comes into the station, if any more in fact do, I will hold and not turn over to the police," McQueen said. He announced that KSAN would carry news coverage of the whole process of deciding whether or not to turn over the material. "We want to keep the whole process of decision-making as public as possible."

Laughlin, however, said on the air that because the station was owned by Metromedia, and Metromedia was a corporation that wanted as few hassles as possible, they would avoid picking fights. The *Barb* reporter commented that throughout the discussion, McQueen seemed intent on defining himself as a professional journalist, and not a revolutionary. At one point, Dick from the *Good Times* turned to McQueen and said:

> "You sound incredibly like a professional journalist."
>
> "He is a professional journalist," put in Travus.
>
> "Ok, that's what you are," Dick went on, "then you're really not one of us."
>
> "If it's a question of..." began McQueen, stopped, and started again, "if it's a question of either I am this or I am not this then, yeah, I'm not one of you."

But the *Barb* reporter would not settle for that and he closed his article saying McQueen was one of them, more than he might know, and if Metromedia decided against fighting hassles, his neck might be on the block.

Duff said he felt strongly about protecting confidential news sources, but he did not feel bomb senders were confidential sources because they sent their letters to be publicized. He felt at the time as he had the previous spring when the station received a letter from the Weather Underground, who claimed to have set off a bomb in Washington, that they should turn such letters over.

O'Hair said the meeting they had on the air with the reporters from the underground papers was something they had to do. They did not, or he did not, really want to, but an underground station could not say no to the underground press. Both he and Duff felt that. Maybe a year ago they would have come out sounding like they were the people's station in the discussion, but this time they came outfront and said they were not.

PART IV:
RENEWAL

Chapter 13

AN INCIDENT
(November 1971-January 1972)

In this incident concerning the obscenity of programming, there is a surprise of expectations about what should be occurring at the station at this time. There is also, in the effort to justify the incident, evidence of conformity with expectations. It is to be noted that both behaviors are characteristic of the station at this time and that the oversight involved, while conspicuous, is not extraordinary.

Nisker was at the station on Sunday night, November 7, the night of the Dr. Hip Pocrates program on which Eugene Schoenfeld had as his guest Margo St. James. Rona Elliott said there were about a dozen people there in the course of the evening. Duff said he was personally hurt that so many of them were there and they did not stop it. It might have cost the company $80,000 in legal fees. It put his career at stake. Duncan said he talked with Duff and Duff understood that. It was not a dirty joke.

There might have been nothing come of it if Warren Sugarman had not written a letter of complaint. But he did, and on Monday, November 8, he sent his letter to the San Francisco office of the FCC with a carbon copy to KSAN. It read:

> Gentlemen:
> I am not in the habit of writing letters of complaint — this is my first. However, I am fed up to the teeth with overpermissiveness and people "doing their own thing" to the point of my discomfort or inconvenience. I am not a prude nor a blue-nose, BUT ———:
> Last night my wife and I retired earlier than usual and decided to try FM radio for a change (and a change it was!). We happened on KSAN-FM and tuned into a talk show format show run by a "Doctor Hip." The subject of the show — with two in-studio guests and countless telephone calls — was

"Oral Sex". The guests on the program were proponents of this practice and spared no details nor restroom wall language in their conversation. I would be more graphic in my recounting of the program, but would probably be subject to arrest.

Suffice it to say that every crude term in the book was on public airwaves. In seventeen years of marriage there have been few times my wife and I have blushed at each other, but this was one of them. We could (and should) have turned off the station, but just couldn't believe what we were hearing and then wanted to find out what kind of station we were hearing it on. I have no opinion one way or the other about the subject of the discussion, but feel that a radio station is hardly the place to missionary for a particular sexual practice and the language was less than fit for public airways.

If it is possible, I would suggest that you procure a tape of Sunday evening from about 9:30 to 10:30 and get an idea of what goes on.

Very truly yours,
Warren A. Sugarman

A week later, on November 15, the engineer in charge of the San Francisco office of the FCC notified Sugarman that he was forwarding his letter of complaint to the local office of the Department of Justice in San Francisco and sending a copy to the FCC in Washington.

On December 15, the chief of the Complaints and Compliance Division of the FCC sent a letter of inquiry to KSAN with a carbon copy to Dougherty requesting a response to the Sugarman complaint and a tape or transcript of the material complained of.

Dougherty replied to the FCC request on December 27 asking for an extension of time for submitting Metromedia's response as he was going on vacation. A few weeks after he returned he submitted a letter defending the objectionable program.

Dougherty said the FCC letter of inquiry of December 15 was not different from other letters of inquiry he periodically received. But he could tell by Sugarman's letter this one was serious. And when he received the inquiry, it was not the first he had heard of it. He got involved after KSAN received their copy of Sugarman's letter and before Duff wrote his reply.

Duff's secretary said when Duff received Sugarman's letter of November 8, he was surprised and he asked her to get the logging tape of the Dr. Hip Pocrates program from Sunday. But when she went to get it out of the studio, she found it was missing and that the logging tape for about four hours of Sunday night's programming was missing. Then

O'Hair wrote up a memo for Duff which recalled the gist of the program:

Memo to: Willis Duff
From: Thom O'Hair, Program Director
Subject: Dr. Hip's Program of 11/7/71

During the opening Dr. Hip makes reference to Paul Krassner's publication of the Rolling Stone papers in the next issue of "The Realist". Theme of the show is "sexuality". An introduction of Margo Saint James is made. They talk about making a good thing better. Margo states that she is a "sexualist" and that she works for pussypower. Margo describes pussy power and the description used the term "snappy pussy". Dr. Hip attempts to describe medically the term "snappy pussy". A description of "snappy pussy" ensues. Cute banter about "snappy pussy". She explains how women can do exercises to enhance their man's pleasure...the conversation mellows out after a while.

Margo tries to move away from the subject after the first break. Dr. Hip returns her to the subject. He asks about the male's part in sexuality. The first caller asks a question about cunnilingus and how to get the fullest of out of the act...

Callers sounded very sincere in their response. Caller changes the subject to a sincere question about his wife taking a psychedelic drug during early pregnancy. Subject then drifts from pregnancy and drugs to a political statement by Dr. Hip. A few more callers and then talk returns to oral sex. Problems were cited, an example, "it smells bad". The term bitch is used but in context. Margo suggests vitamins and yellow vegetables for possible infection and regular bathing. She said, "I must admit that eating pussy is easier than sucking cock". Dr. Hip defines the term as a penis. She goes ahead to explain that some women feel that it is unclean, or not sure it is right — normal.

Caller changes the subject to ask if the use of synthetic vitamins will cause problems of accumulation of harmful chemicals in the body. A caller asks what exercises she might use to keep from getting tired while sucking off a man. Margo explains some exercises she might do.

Duff included O'Hair's memo along with a memo he sent to Dougherty on December 16 describing the circumstances of the program and justifying it as relevant to the segment of listeners to which it was directed.

Duff had been out of town the night the Dr. Hip Procrates program of November 7 ran and he had not heard it. Duncan said Duff was in New York with him. Duff's secretary was in town but she had not heard it either. After she found the tape was missing, she called her friends in Marin County but none of them had heard it. For a while it seemed nobody had. The missing tape became an issue after Dougherty called. He was freaking out. He asked what in hell was going on and told them to get him a tape. He also asked for biographical information on Schoenfeld, the Dr. Hip Pocrates of the program. She had to go out to Stinson Beach in Marin where Schoenfeld lived to get it and Shoenfeld gave her, in addition, a copy of his book endorsed to Dougherty. The book was a paperback published by Grove Press drawing on his Berkeley *Barb* columns. Duff was having a fit.

Krassner of the *Realist* had been on the air before Schoenfeld that Sunday night. Krassner had been fired the week before from KSFX, where he was doing a talk show, and had asked Duff for time on the air on KSAN to tell his side of the story. Duff had agreed to give him time from 6 p.m.-9 p.m. Sunday night. Krassner on the air had read several letters said to be stolen from the files of *Rolling Stone* which had to do with how the *Stone* was exploiting youth markets for advertising. Someone from the *Stone* called afterward and asked for a copy of Krassner's program. Fong-Torres said they want to know at the *Stone* what had been stolen. Mike Hester, a trainee working in production and new at the station, said he took the master logging tape from the studio and made a dub of about an hour and a half of Krassner's show at Duff's request. He then put the logging tape back in a box in the studio where it was kept. The FCC inquiry came after the *Stone* call, Hester thought it was two or three days after, and they could not find the tape.

In an affidavit he signed on February 14 for Dougherty's submission to the FCC, Hester said it was possible he had put the logging tape back on the recorder in the studio where it would have been reused and erased, rather than replacing it in its proper place as he normally would have done. It was possible, and that would account for the tape's being lost. He said in the affidavit his memory was unclear as to exactly what he did with the master aircheck logging tape when he was through with it. Later he said he thought at the time they wanted someone to lay the blame on, so why not him? But his belief was the tape had been stolen.

Duff said in an affidavit he signed on January 14 that the master logging tape had been lost but there was a second tape made each week of the Dr. Hip Pocrates program. When he learned that the master logging tape was missing for the time in question, he obtained the second tape from Rona Elliott who got it as a dub for use in producing the Dr. Hip Pocrates program for syndication.

Elliott said she had done on November 7 what she usually did which was to get a dub of the Dr. Hip Pocrates program from Simmons right afterward. She then took it home and started editing it into about half a dozen shorter reels of five or 10 minutes each, forming weekly features to send to the five station's on Schoenfeld's syndication list. The November 7 program was number nine of his series. The whole show, she thought, had lasted for two or two and half hours. Her edited tapes lost some of the time and changed the original order.

Duff said in his affidavit of January 14, the tapes he obtained from Elliott had been edited and rearranged and placed on several reels. But he was assured by Elliott that they were complete, with the exceptions of certain material edited or destroyed. The edited portions were "er's", "ah's", station identifications, repetitions, and commercials, and not of a substantive nature.

Duff's secretary said Doughtety when he called after getting the FCC inquiry told Duff he wanted a transcript of the program in Washington within a week. It was after his call that they got Elliott's tape and found it was seven or eight little reels. It was bits and pieces of the program. She tried to find someone who had heard the show Sunday night to help them reconstruct the order. That was when she called her friends and it seemed no one had heard it. Then she and a friend of hers named Ted and Duff and a friend of his sat down one night at the station with Elliott's little reels and each of them typed out one or two. That was in the second week after Doughterty's call. Then they got together and put it together, they pieced it together the best they could. Then they timed it and they found they were missing a lot of time, maybe an hour, it seemed at first it was two hours. In the end maybe they lost an hour. They felt they could account for it by giving leeway for commercials, dead air and Schoenfeld speaking slowly. If a question came up they said they would say, Schoenfeld speaks slowly. They had a rough-typed version that night. The next day another secretary retyped it and they standardized the language because each of them transcribing had done it a little differently. Where one of them had typed "ah", another one might have typed "oh" or "ahh" or "uh."

Duff in his affidavit of January 14 said he believed the reconstructed transcript accurately represented the Dr. Hip Pocrates program of November 7, with the exception of Elliott's editing and some special difficulty in reconstructing the original order of the last half, but that these exceptions were of an inconsequential nature.

Duff's secretary said after they had the transcript typed up she made 20 copies and sent two or three to Dougherty in Washington and bound one in a book she titled "Sex and Stereo." Some of the copies circulated among the staff and one she used to refer to when she called people for letters in support of Schoenfeld for Dougherty to send to the FCC. The letters testified to Schoenfeld's competence and to his having performed a local public service in dealing with oral sex so frankly on his November 7 show.

Duff said when he got back from New York two days after the show, he had to impress the staff that it was serious. There were two things that happened that Sunday night that should not have happened. Only one was the subject of the inquiry by the FCC. O'Hair said the FCC would really freak if the other got out. St. James, the guest on Schoenfeld's show, had come in earlier in the evening while Krassner was on and had given Krassner a blow job in the studio. St. James was an old friend of Krassner's. She was the *Realist* nun in his newspaper. Elliott said the real problem was that Krassner could not get a hard on. Duff said when he heard the tape of Krassner's show that they had dubbed for the *Stone*, he could hear the blow job, you could tell if you were listening for it.

The FCC had no specific rule prohibiting sexual relations in broadcast studios but alcohol was prohibited and the import of regulations on obscenity and profanity was that a broadcaster should have knowledge and control of what went on the air. Duff as station manager was the man considered most responsible in the Dr. Hip case. Duncan said even though Duff was in New York at the time, the radio station was his responsibility. Duncan said personally he thought the Dr. Hip program they ran that night was a mistake, but they could not publicly admit that. Publicly they had to defend their responsibility as broadcasters. Radio was not like a newspaper or a magazine, it was not a medium of choice, it did not allow you privacy. Once you turned it on, you were subject to it. The broadcaster had to be concerned about social mores and good taste.

Duncan said personally he felt radio was not a vehicle for testing obscenity codes, but he knew there were people in the FCC and the Department of Justice who wanted to use it for that. For him there were two issues. They were not the FCC's issues but internal problems so far as the Radio Division was concerned. One was whether the Dr. Hip program had been in bad taste. The other was whether the general manager and program director had been in control of the station.

Duncan said for him the prime issue was controls. That was the basis for the FCC's granting the license. But he could see how for Dougherty the issue was different. For Dougherty it was the programming content. Dougherty knew the FCC was looking for a test case of the obscenity laws and he did not want this to be that case. For Duff, Duncan thought, the issue was different. For him it was that the staff there had betrayed him. They had not taken seriously the trust of the station while he was gone. If they had they would have controlled it and not let either of those two incidents occur.

Sunday nights at KSAN had been relatively loose nights in terms of programming for at least a year prior to November 1971. Live music shows had been on since the spring before the Dr. Hip Pocrates program. Up until the end of October, Laughlin had been on with his talk show right after the live shows. Like Schoenfeld, Laughlin had guests on his show, their conversation was not prerecorded, there were listener call-ins, and you could not be sure in advance how it would go. Elliott said she was at the station the night of November 7 screening incoming phone calls for Schoenfeld and she did not find them objectionable or feel obliged to stop what was going on. Simmons was also there as the on-air producer and she did not stop it. St. James was there as the guest. Schoenfeld was there, and Krassner. Nisker was there for a while, Bear was there with his old lady Goldie, and there were some others who came later. There were no complaints called in during the show. Afterward Elliott got her tapes and took them home and started editing them.

The following week, O'Hair called her and told her he did not want that woman, St. James, coming into the station ever again. He told her he had been told that she, Elliott, had set it up. He accused her of having a low moral character and said he did not want any of her friends around the station. Duff called her after O'Hair and said he had heard she set it up. Thinking back on it, Elliott felt they needed a scapegoat and she was a good one because she did not work for Metromedia. They were pissed off at Krassner for having got head. The FCC letter of inquiry had not come yet. There was no Dr. Hip Pocrates program on the air the next week.

Elliott said Duff when he called asked if she could be bribed or if he could take her to lunch for her copy of the tape. He said he would lose his job if this did not come through. She gave it to him and he never gave it back. At one point she felt they were trying to discredit St. James. Duff's secretary called and asked if St. James was still a hooker. They knew she used to be one. Elliott took offense at that. St. James was her friend and so was Krassner. St. James was one of the finest people she knew. St. James believed in pleasure, she was what you might call outrageous, she was 34 and she lived in a house in the woods in Marin.

Schoenfeld said the week before his November 7 show, on Sunday, October 31, Halloween night, he had a practical therapist on as his guest. St. James walked into the studio and started fooling around with him. She was dressed as a nun. After the show he went with her over to KSFX where Krassner had a talk show on and St. James gave some advice to the audience. She said everyone should get head. That was the kind of thing she did. Krassner was fired from KSFX the following week. After he was fired he asked Duff if he could come on KSAN and explain what had happened.

O'Hair said he was opposed to Duff's hiring Krassner but Duff wanted him. Duff was a *Realist* freak.

Elliott was 24. She had worked for a while at KMPX as a secretary when National Science owned it. She felt Schoenfeld's Dr. Hip Pocrates program on KSAN was a public service. It really helped people and it gave them a lot of laughs. She and her friend Sally did the syndication. Personally, she felt doing the show helped her. She saw she could really help someone else, that people could talk about cocks and cunts, people could see that everyone had them and that was where it was at. They could see they were not the only one with a problem. The November 7 Dr. Hip Pocrates program seemed to her the best of the nine they had done to date.

The transcript of the program that was finally put together from Elliott's reels of tape ran 55 pages. Page one had Schoenfeld's introduction:

> Doctor: Last week we were speaking with a lady from a Berkeley clinic for treating sexual problems.... This week, I thought we would talk not about sexual hangups but about, more or less, normal sexuality — whatever that is — or at least I want to get into areas where people are happy, more or less, in their sexual roles, in their sexual developments; but we're going to talk with Margo about ways of improving — making it a bit better. I guess that was the way of capsulizing what we're going to be talking about: making good things better.
>
> The phone number here is 986-6244, and a little later we'll be answering your phone calls.
>
> Margo, can you tell us what you are — you have something to do with exercise and I wonder if you could tell the listeners what kinds of exercises and how long you have been doing this. Maybe, by way of introduction, I should say that Margo was the Realist nun — those of you who have been following the Realist — and she used to do things that, then, were thought of as outrageous. I guess people would still be outraged by a nun seeing someone off at an airport and suddenly falling into a heavy clinch.

	Margo, what — you say you're an exercise specialist — what was the title you told me?
Guest:	I said 'sexercise' and ———
Doctor:	Actually, there is a book called "Sexercise" and it's not bad.
Guest:	I'm just stealing their word.
Doctor:	You're a sexercise therapist, is that right?
Guest:	You could — well, I'm into, I work for pussy power.
Doctor:	Pussy power. What is, ah, how would you define that?
Guest:	Pussy power is women's sexual potential. It's infinite actually, I think.
Doctor:	Yeah. Well, what's wrong? Don't you think this potential isn't being exercised, so to speak?
Guest:	True, yeah. The frozen pelvis abounds in this country; and they shed their girdles, this last generation, and the women's muscles are still unworked. Very few women have what men refer to as "snappy pussies". What do you call 'em on this show, anyway?
Doctor:	Well, on this show we'd probably say, we'd probably say that you believe that most women don't have the full development of muscles which surround the vagina.

On Thursday, November 11, Caen had an item in his column in the *Chronicle*: "The brass at KSAN-FM has its fingers crossed after Dr. Hip Pocrates on that station Sunday. Miss St. James, rising to a new low, out-Lennied the late Mr. Bruce to the point where there have been complaints to the FCC from outraged listeners."

Schoenfeld said after Duff got Sugarman's letter of complaint, he was not sure if it was after the letter of inquiry came from the FCC or before, Duff called him into his office. Duff was concerned about Krassner's show and the blow job the most but was also upset about his program. He banned St. James and Krassner both from the station and asked for letters of support for submission to the FCC. Later he asked for more letters. They had several meetings. Duff wanted documentation.

Schoenfeld said he got the impression that Duff was more than protecting the station, he was protecting himself, his reputation. At one point he and Duff and O'Hair and Elliott all got together and listened to the tape of the show and they were all laughing. Duff asked him in the future to leave a memo each week to make sure Simmons knew who could be his guests and their topics. Schoenfeld said so far as he knew, the FCC complaint issue was whether the station manager had control of the station. He could see there was no point in doing a radio show that was going to be thrown off the air. Also he was concerned that Duff would lose his job or that Simmons would be hurt.

It was Simmons they finally got after for not having used the kill switch and excised the objectionable language. But no one was especially convinced as to when using the kill switch would have made a difference. Maybe the only way to have saved it would have been if they had stopped the whole show. Simmons said she did not want to talk about it. She felt she should never have come in that night. But she did give a deposition on January 6 for Dougherty to use for the FCC:

(1) My name is Bonnie Gail Simmons. I reside at 290 Marion, Mill Valley, California.

(2) I am twenty-three (23) years of age. My education includes three years of college at the University of Colorado and the University of Arizona, where I majored in fine arts. My familiarity with the radio business has been broadened by virtue of the fact that my husband, Robert Simmons, has been a professional broadcaster for more than four years and is currently an employee of The American Broadcasting Company at Radio Station KSFX in San Francisco. My prior work experience was in the related fields of the recording and entertainment industries. My experience and training has familiarized me with the general rules of operation in commercial radio.

(3) Among my duties at KSAN, I am the producer of the "Dr. Hip Pocrates" program with Dr. Eugene Schoenfeld. Preparatory to becoming a talk-show producer, I trained for four weeks under the guidance of Peter Laufer, the chief producer of talk shows at KSAN, and functioned as producer of another talk-show, the "Travis T. Hipp" program before assuming the duties of the "Dr. Hip Pocrates" program.

(4) KSAN Utilizes a seven-second delay device, and among my duties is control of the "kill" switch. I regularly make the decisions as to when the switch should be used.

(5) On the night of the seventh (7th) of November, during the program in question, several exceptional circumstances existed. Normally the program immediately preceding the "Dr. Hip Pocrates" program is a live music show, requiring little more than maintaining the levels. On the seventh,

however, that three hours was used for another talk-show which I produced. Also, that evening I was not feeling well, due to an abessed tooth which became progressively more painful as the evening wore on. Although I took the prescribed medication, the discomfort was still distracting. Also, the night of the seventh, due to repairs being done on our main studio, we were originating the program from an auxiliary studio, which is not ideal for the production of a talk program.

(6) During the course of the program, it was my opinion that the content of the show was proper and should not have been cut. I am normally very conscientious about cutting objectionable language, and although I was moderately uncomfortable about the language used during the course of the program, others were present in the studio at the time and no one seemed to be in the least uncomfortable about the language, and this may have influenced my decision not to use the "kill" switch. It was not until several days later that I felt certain portions of the program should have been cut.

Dougherty said he felt the FCC and the Justice Department were looking for a test case of the obscenity laws to determine what constituted obscenity. There had been a prior case, Eastern Education Radio, concerning an FM station in Philadelphia, where the Commission had made the fine small, $100, hoping the stating would test it, but they did not. They did not have the money. Metromedia's fine for KSAN for the Dr. Hip Pocrates program could have been $10,000 in the criminal code. Dougherty said he felt at the time the FCC was looking for someone who had the money and the legal resources. But he did not want to be the test because of the facts of the case. St. James was not a doctor and there were other facts. It was a bad case. Bad cases made bad laws. They set bad precedents for other stations. In defending KSAN, he was protecting more than the station. He was protecting other stations and trying to keep from saddling the industry with a bad precedent.

Dougherty said his object in drafting his response to the FCC was to frame it in such a manner that they could not push it. He wanted to show free speech but that one girl had one lapse. Schoenfeld had qualifications. The subject was all right. They had letters from people in the community. They acknowledged they had solicited the letters. He wanted to cover it on all grounds, obscenity was only one. He wanted to be ready for both the FCC and the Justice Department. He felt the community standards that existed in San Francisco were a lot different from the standards that existed in the minds of the regulators. It was not going to be easy. There was a question as to St. James' qualifications as an expert. The discussion they had was far from clinical, it was pretty goddamn earthy. Duff's career was at stake.

Dougherty said Duff told him he felt his career was at stake and he told Duff he knew it was, that there was a threat from overhead. The whole thing was extremely serious because it could have involved the license, criminal prosecution for obscenity, a fine, and there was the question of what the hell had happened when they thought they had the situation there under control. Well they did have it under control, but there was a judgmental lapse.

Dougherty prepared two submissions in response to the December 15 inquiry of the FCC. The first, dated January 17, was a letter of defense with attachments: affidavits from people involved suggesting they had known what they were doing, a transcript of the program of November 7, letters of support for Schoenfeld, and biographical information on Schoenfeld and St. James. The second submission included additional letters of support and a copy of an article from the San Francisco *Examiner* of February 2, "Dirty Words Come Clean." Dougherty said they put together the first submission in Los Angeles. He went out there and queried the people involved pretty goddman well. He had to believe he was telling the truth. The letter of defense in the first submission ran to nine pages. It was addressed to William B. Ray, Chief, Complaints and Compliance Division, FCC:

> Dear Mr. Ray:
> This will refer to your letter dated December 15, 1971, which forwarded a complaint that had been received by the Commission concerning the operation of Station KSAN(FM), San Francisco, California. The complaint related to the broadcast of the Dr. HIPpocrates program on November 7... .
>
> A transcript of the Dr. HIPpocrates program appears as Attachment No. 3 to the First Affidavit of Willis Duff, general manager of the station. Mr. Duff's affidavit also describes the circumstances surrounding the broadcast and goes into other areas which will be touched upon later in this letter.
>
> Before doing so, however, it is necessary to make reference to the Second Affidavit of Mr. Duff. As indicated therein, duplication of the transcript of the program was assigned to Mike Hester, who is a trainee at KSAN under the United States Army program, Project Transition. For some unknown reason, he neglected to "dupe" the whole program and then misplaced the master aircheck tape. As a consequence, the second half of the transcript had to be procured from other sources. Nevertheless, the transcript contained in Attachment No. 3 of the First Duff Affidavit is believed to be an entirely accurate representation of the substance of the program aired.

As indicated in Duff's First Affidavit, the "Doctor HIPpocrates Porgram" was conceived to fill a need in the community for a discussion of public and individual health questions. The program is aired Sunday evenings, commencing at 9:00 p.m. It is aimed at the young adult, particularly those of the genre that might be categorized as the "disaffiliated". Rating information discloses that the demographics of its audience is almost wholly adult.

Dougherty went on to speak of the redeeming social value of the Dr. Hip Pocrates program in the San Francisco community:

The "Dr. HIPpocrates" series has been particularly well received in the San Francisco community, and it has evoked very favorable response from members of the medical profession and laymen alike. See Attachment No. 6 to Duff's First Affidavit. To the best of our knowledge, no complaints, other than that lodged by Mr. Sugarman, have been submitted to the Commission with regard to the program aired November 7 or any other program in the series.

Admittedly, the subject matter of the program — oral sex — falls in a very sensitive area. But this fact, standing alone, should not necessarily rule the subject beyond the bounds of discussion... .

Perhaps the subject could have been discussed by the doctor in a clinical fashion, where he and his guest utilized the legal terminologies for the matter under discussion. But this then raises the question as to whether the program would have fulfilled any need, since the discussion may not have been understood by many members of the target audience to which it was directed. In this connection, it should be noted that Mr. Duff remonstrated the producer, Mrs. Bonnie Simmons, that she had been remiss in not excising the common vulgarities (she did kill certain words as the transcripts indicates)... .

The program was not aired to shock or titilate the audience. Quite the contrary; it was aired to fulfill a need for a discussion of an adult problem. See Attachment 3, page 44, which indicates that one of the most pressing problems of adulthood is sex but there is a reluctance to pursue its discussion.

Dougherty next referred to the case of Eastern Education Radio, arguing that the rule in Eastern did not apply:

From a legal standpoint, the question involved here is whether the single broadcast in question involves a transgression of 18 U.S.C. S1464. Based upon Eastern Education Radio, 18 Pike & Fischer R. R. 2d 860 (1970), the broadcast in question did not involve anything that could be characterized as obscene. Further, there are so many distinguishing features between the broadcast involved here and that involved in Eastern, that the rule of that case does not apply.

First, the broadcast in Eastern was taped and there was ample time to edit the program and to bring it to management's attention to determine whether it applied to extant policy statements. Second, the speech involved in Eastern had no redeeming social value and it was patently offensive by contemporary community standards. The discussion involved here had a redeeming social value and, as correspondence contained in Attachment No. 6 to the First Affidavit discloses, the speech here was not offensive by contemporary standards in the San Francisco community. Third, the speech in Eastern was gratuitous. That was not the case here.

Toward the end of his letter, Dougherty said what the Commission ultimately had to decide was whether the program in question was indecent by contemporary community standards existing in San Francisco, not in Washington, D.C., and whether it was completely devoid of redeeming social value. An important point, he felt, aside from the threat to free speech that a negative ruling night imply, was that the Commission was here intrepreting a criminal statute and language of the statute had to be strictly construed. As the Commission had stated in Eastern: "The Commission can appropriately act only in clear-cut, flagrant cases; doubtful or close cases are clearly to be resolved in the licensee's favor."

Chapter 14

KSAN
(November 1971-April 1972)

During this period occasions for terminating special programs are opportunities for making statements about criteria for what is now appropriate to the station. Most obvious are criteria of improving quality, broadening audience, and avoiding trouble. A persistent dilemma suggested by criticism of the station at this time is that of isolation. Internal distinctions made by the staff are often not comprehended in assessments made by others. One result is that the station seems, and, is to a large extent inaccessible, but for different reasons now than earlier in its history.

Duncan said Duff was the kind of guy who only made a mistake once and that after the Dr. Hip inquiry, Duff tightened up at the station. McClay said the Dr. Hip thing made everyone at the station paranoid. He used to do "come" sets, but after Dr. Hip, Duff got touchy about their playing certain records. McClay said it seemed to him the only time they got in trouble was because of the talk shows and then the music people ended up paying for it. He thought Duff got uptight because the loss of the tape might look like a coverup to the FCC. Ponek saved a memo he got from Duff on February 2. He said he felt Duff would not have sent one like it before Dr. Hip. The memo read: "Stef: I cannot understand your playing 'Sammy's Song' under the current situation. That was a real lapse in judgment. That cut is definitely a no-play. WD."

O'Hair said he felt the Dr. Hip thing really made Duff scared for his job, because of the Krassner show before, and that Duff put himself into handling it discreetely and began looking for another job. Harris said he felt the Dr. Hip show was the undoing of Duff. Duff spent the rest of his time with the station fighting it. Metromedia could not fire him because they were threatened with a loss of license. Duncan said he felt the FCC in granting the license had granted them responsibility for control over the product. That people were not released, that no one was fired, showed the company felt controls were implemented.

Duff said he felt they did well. It was a stroke on behalf of free speech. Most of it was between himself and Dougherty, although some of it got to be the talk of the company. But he felt in the best of graces with the company when he left. Duff left six months later, in May, to go to work for a smaller firm a friend of his had started. He told Duncan he was leaving 30 days before and recommended that Donahue replace him as general manager.

Donahue said he felt Duff was afraid for his job after Dr. Hip. Duff had not turned the station around to earning money yet. But Dougherty was smart in handling it, he expected the FCC did not want to go to court over an obscenity charge because of the other media. Donahue felt Dougherty protected the jobs of the people at the station in his defense of the program. Dougherty knew they could not fire Simmons or Schoenfeld or it would look like they had been wrong. Donahue said personally he thought Schoenfeld was a bore on the air, and that was not a reflection on content. He told Duff when Duff wanted to let Krassner on not to, because Krassner did controversial things, but Duff let him on anyway. So far as Donahue knew, the handling of the Dr. Hip case was kept secret in the company, it was an insider's dirty joke to only a few.

Sanders felt the real trouble in the fall and winter of 1971 did not have to do with the Dr. Hip Pocrates incident, but with the general philosophy of people at the station, and she felt she knew because for a while she had it. She had been on the air weekends and doing fill-ins in addition to being traffic director from August up until January, and she felt that during that time she joined the clique in-group at the station and adopted their philosophy of deceit, that they would say this but really be doing that. They would be out for themselves, ego tripping, but saying they were in it for the station. To get what you wanted from them, you had to act like them and think that way, have egomania. Sanders felt for a while she did, she was swearing like them and speaking her mind, putting on a show a lot. It nearly drove her old man crazy. He was a student at the California College of Arts and Crafts in Oakland. They had been married seven years. He did the show with her sometimes. She thought of it as a Tom and Raechel, the Pam and Bob Sanders show, but with country music and in a more laid back, less commercial style.

Sanders was taken off the air in January and Hester was put in her place. Hester was black. Sanders said she thought it was a political move. Hester said it was because Sanders' show was just no good, the ratings were down. Sanders said she felt it was one thing to put on a show and be what you were on the air, which was what she wanted to do. It was another to feel you were different on the air than you really were, and different meant not only saying one thing and doing another but really being another way. That was what got to her about it. She could see how it happened with Street.

Street was back on the air in November, having taken off in October for a rest. The sales gross for the month was the highest in the station's history. It was up 36% from October and up 71% from November of the year before. Skinner said in his sales report the station was sold out Wednesday through Saturday in just about every time period for most of the month but that was all right. It was pre-Christmas again and Gardner came up with the idea of giving away trees planted at the Point Reyes National Seashore as Christmas gifts to clients. Skinner said he thought it was a good idea instead of the usual coffee mugs and calendars. They had arranged a trade deal for 500 trees.

The main studio was in full operation by the end of November, with the remodelling done as proposed by Kibler, but with a few bugs still to get out. Duff said in his monthly report the air staff experienced a resurgence in morale and the sound of the station was good. O'Hair said:

> Well, we finally purged ourselves of October's flu, ingrown toenails, toothaches, and insecticide poisoning, and this month everyone was in good spirits and made it in most of the time.
>
> Tip to all you Metromedia Program Directors: To ward away sickness — stand on your desk, clad only in your shorts, with a roman candle in each hand, chanting "Eighteen Hammers" and "Oh Susannah" four times, put one APC tablet under each turntable and stay away from all contact sports for 24 hours.

O'Hair said also in his report that on Thanksgiving Day, Ponek and Laufer called Plymouth, Massachusetts and spoke with the president of the American Turkey Association. They called a bar in Provincetown and spoke with several Texans who were reported to eat armadillo for Thanksgiving, and they spent time on the phones, locally, hooking up people with extra turkey with people wanting somewhere to eat.

Duff and Ponek received awards in November at Bill Gavin's Sixth Annual Radio Programming Conference. Duff was named station manager of the year for progressive or free form FM, Ponek was named disc jockey of the year in the same category. Ponek said he had not entered the Gavin contest himself, although he was asked. Duff had sent in a tape for him. But getting the award was meaningful. It was an establishment prize and it represented a lot of things the station was against. But he was not so snobbish that it did not matter. It made him feel good. It was recognition by his peers.

In December, Duff fired Cole, then changed his mind and took him back. Campbell felt that was typical of Duff. He could not fire anyone except in anger and then when he did the staff could manipulate him and make him change his decision. The staff had a meeting after Duff fired Cole. Campbell said she saw Duff walking back to his office when it was over. She asked him if he had let the staff manipulate him again. He hung his head and said something implying he had. In her eyes, that was weakness. In Gossett's eyes, it was not. What mattered to Gossett was that Cole be allowed to stay. Cole was his friend and he felt what happened to Cole could have happened to him or to any of the air staff.

From what Gossett knew, Cole had left the transmitter unattended the night before and there had been about 10 minutes of silence on the air. Duff fired Cole when he came into work the next day. The staff called a meeting to protest it. They told Duff at the meeting that it could have happened to any of them.

McClay said he felt it was Bear's fuck up, not Cole's. Bear had a habit of lateness and no one had his phone number. Cole was new and maybe he did not know what to do.

Cole said what happened was he was doing Street's shift that night, it was a Wednesday. He was going to split at 2 a.m. as soon as it was over. When the time came, Bear who was supposed to be on next did not show. His relationship with Bear was awkward but he played an opening for Bear and then he played records for him for about half an hour. He tried to get Duff on the phone but Duff was out of town. He tried to get O'Hair but he could not get him or anyone. Then he got a call from Bear's old lady saying Bear was on his way. After that he called Kibler and Kibler said to turn the transmitter off. He did not do that but put on a long record, the Environment's record of "The Sea." He put it on at 2:30 and then split and went home. At 3:30, when he got home, he turned his radio on and Bear was there.

The next day when he came into work, he got a note from Duff's secretary saying Duff wanted to talk to him. Bear had complained about his unprofessional attitude in leaving the transmitter unattended, and also it was illegal. He met with Duff and Duff said he would have to let him go, he had talked to attorneys. O'Hair said he had been in radio a long time and he had never heard of anybody doing that. Cole said he had. He walked back into the record library to Simmons. A lot of people were around talking about it, they were in support of him and they wanted to have a meeting. They had a meeting and Duff came and they told him what he did was unjust, it could have happened to anyone. Duff agreed to take another 24 hours to consider it. The next day, Duff called Cole into his office and said he was upset that Cole had rallied the troops like that. Cole told him he had not rallied them. Duff said he had reconsidered and he was not going to fire him but he was going to suspend him for a month. Then they had a meeting and Duff announced it to the staff.

Duff said what happened was Cole left the transmitter unattended for 20-30 minutes. He got furious at Cole because it was negligence and he took it as a personal betrayal. Then the staff organized and destroyed that as the issue.

O'Hair said what made the difference was that he organized the staff. They did not just do it themselves. He deliberately seized the opportunity of Cole's firing and used the staff to challenge Duff and show Duff that he was not going to be Duff's flunky. He felt Duff had made the decision to fire Cole out of rage and not reason, like a father punishing a bad child. He thought Duff saw when he started that he was going to organize the staff against it and that Duff was surprised that he was on the side of the staff. He wanted Duff to see that he could manipulate the staff better than he could.

O'Hair said he felt the Cole firing incident showed a change in his relation to the staff. Whereas before, ever since he came on as program director in August, the staff was testing him and evaluating him behind his back, now, he felt, in standing up for them to Duff, it showed he had gotten acceptance and respect from them. And that was aside from the fact that he expected if ever he was in the position of station manager like Duff and anyone left the transmitter unattended, he would probably do what Duff did, fire them, and yank their license.

On Friday, March 3, at about 3 p.m., KMPX ceased to broadcast rock music. Sanders said O'Hair and Duff laughed when they heard and gloated, but she felt sad. Bear said he felt sad too. They had not been competitors really. KSAN was always the best. KMPX was their brothers and sisters. John Wasserman, who had replaced Gleason weekdays in the *Chronicle*, wrote in his column on March 5 that as of 3 p.m. Friday, March 3, KMPX, the old KMPX, had died. As of 6 a.m. Saturday morning, a new KMPX was born. The old KMPX had featured rock music and depending on your predilection for labels was either underground or free form. The new KMPX was middle-of-the-road, an easy-listening station: "KMPX is dead, long live KMPX. The world's original first and only underground rock and roll music FM radio station, the world's oldest permanent free-floating format, has gone defunct."

An article in the *Examiner* of March 4 said it was 3:10 p.m. on Friday, March 3 that the old KMPX had gone off the air and that it had gone off abruptly to thwart the anger of some of the staff. There was a station manager, a woman, who was staying on despite the format change. She said five of the air staff had planned to express their candid opinions of the oncoming "harp music" during the dying hours of the rock format so the management cut it off. The *Examiner* reported one of the outgoing disc jockeys said a security guard had appeared on the premises Friday afternoon to keep the staff from walking off with a transmitter or something, and that some of the staff felt the change to Mantovani was really a ploy of management to prevent them from organizing and getting better pay and working conditions as they had been trying to do. The station manager said the five discharged rock music air staff were given auditions to determine if they could continue with the new format. They were given scripts to read and told the kind of voice wanted was warm and lively, but not top-40 and not far out. She would not say how they did in the auditions but she said she had been aware of the impending change in the format for several months, ever since it became obvious that the station was not going to succeed commercially against its chief rival, KSAN. The staff, however, had not been told of the change until this week.

The *Examiner* mentioned that the KMPX program director at the time of the change was Reno Nevada. He had originally worked at KMPX as a part-time disc jockey under Donahue and had come over to KSAN as a weekend man after the strike. KMPX had gone through several turnovers of staff since then and had staff-management conflicts repeatedly, the most dramatic occurring in October 1970 when the Collective locked themselves into the studio in defiance of policies of the National Science Network. At the time of the format change in March, National Science still owner KMPX. Wasserman, in his column of March 6, quoted Donahue as offering what might be considered an epitaph for the station: "The music was always good, some great people were involved, and the management was at different levels of rotten."

The *Good Times* of March 10 carried an article on the demise saying KMPX, the granddaddy of underground rock radio, was now the illegitimate child of middle-American music. The *Good Times* in the same issue ran an article on KEMO-TV, a UHF color television station also carried on San Francisco cable, which Crosby had bought and started operating in February. Crosby's style of programming KEMO-TV was like the one he had going on KMPX just prior to the 1967 format change to rock, with time brokered out to independent producers and foreign language programs accounting for much of the income. The *Good Times* reported the price per hour for a one-shot deal on KEMO-TV during prime time was $485. On an ongoing basis, prime time would be $375 an hour, roughly comparable to the cost of putting out an underground newspaper.

Crosby had hopes people would go on the air on KEMO-TV with enthusiasm like they had on KMPX, but this time not giving him trouble. Already on Saturdays there was a live show of which the *Good Times* described as reminiscent of American Bandstand. The Congress of Wonders had been on, so had a North Beach street artist and some other unexposed talent. The *Good Times* referred to one recent Saturday night show as indicative of the whole operation: it was so poorly conceived and executed it was embarrassing. But KEMO-TV should not be written off on the basis of its first few fumblings. The important thing, the *Good Times* said, was that it was happening at all, and soon there might be a sorting out of talent down to some really capable people.

Crosby said after he sold KMPX to National Science, and he really hated to sell the station, he only got half of what it was worth, he went away to recover. When he came back he tried to buy other stations, one radio station in Salinas and one in Washington state. He tried to buy channel 38, a television station in San Francisco, but that fell through. Finally he bought KEMO, his ex-wife loaned him the money.

The *Stone* of March 30 carried an item in Random Notes on the death of KMPX and immediately following it an item on Crosby and KEMO-TV:

> Farewell to Waterbeds: KMPX, the first 'hippie FM' station, is finally dead; gone to a kind of muddled, middle-of-the-road format... Meanwhile: Leon Crosby is owner of KEMO-TV, Channel 20, hoping to begin nightly chunks of "underground programming" within the month. Meanwhile: Larry (The Lion) Ickes, one of the Scab Era KMPX men, is over at KFOG-FM, spinning Mantovani on the housewife shift. And meanwhile: KSFX, among other ABC-rockers, has gone to a programmed quasi-playist and cut DJ talk to 15 seconds between sets. And how are things in your town?

Billboard of March 18 reported the KMPX format change, as did *Variety* on April 5. The *Variety* article had as part of its headline the question, "Is underground radio a bad trip?" In the answer was a quote from O'Hair of KSAN on why KSAN was successfully surviving while KMPX was dead. O'Hair responded to a statement made by the KMPX station manager attributing the success of stations like KSAN and WNEW-FM to their commercialism. He said he felt it was more than that:

> "Of course we're commercial! Either you're commercial or you're listener supported — it's either fish or fowl... We accept Modess ads so we

can do a documentary on George Jackson." If KSAN and other successful underground stations have changed, he continued, it's because "society has changed since '68 and '69 — since that summer of love when the revolution seemed just a joint away."

O'Hair added that KMPX's problems could be summed up by saying that it broadcast to a now non-existent audience. Today's underground radio must be less political, not because of "establishment" pressure, he contended, but because its audience has been frustrated out of radical fervor. The major part of KSAN's audience, he continued, is not typified by the activist or "street freak," but by "the secretary in San Mateo who sneaks a joint on Friday nights."

O'Hair noted that the Bay Area market has been flooded with rock stations — ranging from Top 40 to acid underground — competing for the same audience. Inevitably the weaker ones will be weeded out, he said. But he added that underground radio can thrive as long as it is willing to compromise as KSAN does. "To compromise isn't to cop out," he said. "And let's face it, a dead radical is either a martyr or a corpse — a dead radio station isn't any better."

Ponek said he felt as far as competition, KMPX was never really a serious threat for long, but they were into experimentation so you had to keep an eye on them, every once in a while you could pick up something from them. Any competition did that, it kept you aware of yourself. The other FM rock station, KSFX, the ABC station, Ponek felt he never took seriously as a threat either. But he could see how KSFX kept them at KSAN from killing themselves. KSFX with their hit list made the staff at KSAN play close to a quotient of hits, and it made them careful knowing people might tune off from commercials and self-indulgence trips and go to KSFX. But Ponek felt KSAN was basically in a different position from KSFX because it was a community radio station, it had been from the beginning, it had roots. It was like KSFO that way but for a whole other generation. Ponek felt if there was something going on that the people that listened to KSAN were interested in and identified with, they knew the station would be in touch with it. If it was not they assumed it should be.

Two subjects of interest on the station in the past year had been gay and women's news programs. At the start of March, both were discontinued. McQueen said he discontinued them because of poor quality. The *Barb* of February 25 suggested he discontinued them because of prejudice and said it was the second time within a year that the station had cut gay people's news. The program that was cut had started in mid-December. It was initially a 10-minute news feature on Saturday afternoons done by a member of the San Francisco Gay Activist Alliance. The previous gay news show had been cut in June after having been on three months. The *Barb* quoted the KSAN newsman Laufer as saying the recently cut show had contained only a calendar of gay events and that its producer had been making slanderous accusations about police departments murdering people that he could not substantiate. Laufer said in the future the station would only carry news it considered important about gay people, this would be on regular newscasts, and he felt it would broaden their response to the gay community.

The *Barb* also mentioned that a women's news show broadcast Monday evenings on KSAN had also been discontinued the previous week. The women's program had been on for two years and was a project of the San Francisco Women's Media group. McQueen said he had worked with the women at the station. At first they did not have technical skills. He had assisted them in making their own studio. But after a while they did not want his help. They considered him a male chauvinist pig. It was agreed in the beginning that the women's program would be on for 30 minutes a week. Then he cut it to 10 minutes. But the shows were still bad technically so finally he discontinued them. He told the women no, and said he would filter their stuff through the regular newscasts. That was how it was with specialized programs. At first he had thought mostly what was involved was giving them air time. But he found out more was required. They had demands for studio space and technical skills, these were demands on the station's limited resources. The station as a whole had become more sophisticated and its quality had gotten higher, it was necessary to sound professional. The women's shows were not. They were still mainly full of rhetoric.

Duff said the gay women's shows were shitty at the beginning. He expected they would improve over time, but they did not improve.

On Wednesday, March 8, about 30 members of the San Francisco Women's Media group and their supporters held a demonstration outside the KSAN studios at 211 Sutter Street in protest of the station's discontinuing their program and objecting generally to the station's sexist hiring policies. It was International Women's Day. The *Chronicle* of March 9 ran a picture of the demonstrators with a caption quoting Duff on his reasons for dropping both the gay and women's news. He said they were dropped because they required too much of his staff's time and because the quality of both was continually unacceptable. With respect to hiring policies, he said the station did have females on the staff and that one was watching the demonstration.

O'Hair said he felt at first Duff was intimated by the gays and the women and would not throw them off the air. Finally he told Duff he thought Duff was a cunt and it was a choice between firing him or offing them. Then they looked for an issue and did it.

Sanders said she took part in the women's demonstration on March 8 and in the women's day programming on the air. When she went on the air she said she had learned a lot at the station and that she had been given breaks by the station and then was able to get job offers from other places, and this was after she had been cut out of her show. Her going on the air, she thought, indicated her identity was still with the station, despite the fact she felt the women at the station all got treated second class, except for Campbell and Street. Downstairs Laufer was trying to interview the protestors for the news but they would not talk to him.

O'Hair said he felt the women came off as fools on the media. They got television coverage of their demonstration. He said he told them before that would happen, that they would come off as fools, but they would not believe him. He saw the demonstration as publicity for the station, he felt they got more audience by offing the women than by having them on, and even before they offed them they were diminishing them. He told Duff he thought it was promotion.

On Saturday, April 8, Donahue ran a KMPX retrospective special on his show on KSAN beginning at 6 p.m. It was five years and a day since he had first gone on the air on KMPX. The *Chronicle* listed among its radio highlights for April 8: "Tom and Raechel Donahue present a special program dedicated to the celebration of the fifth anniversary of underground radio (6:00, KSAN, 94.9)." The special as it ran included tapes of shows and music from KMPX during 1967-1968, tapes of a few of the foreign language programs which had taken a while to be edged out, and comments by people who had worked at the station recalling what it was like. They stopped in at the KSAN studio Saturday night to reminisce. Some of the people who had worked at KMPX were now with KSAN, some were at other stations, some were no longer in radio. Of the KSAN air staff, three of the five weekday jocks had worked at KMPX back in 1967-1968: McClay, Street, and Bear. Boucher, still at KSAN as production director, had done the commercials and been chief engineer at KMPX. Voco, who now worked weekends on KSAN, had done air shifts on KMPX. Laughlin, who did the KSAN talk show and sometimes sold time, had been a salesman for KMPX. Donahue had been program director. Hamilton had been his assistant and had handled the business of North Beach Productions which carried part of the KMPX payroll.

Nemerovski, who had not been with KMPX, sold sponsorship of the retrospective to Leopold's Records of Berkeley. Dymond, who also had not been at KMPX, said in a press release she sent out that Leopold's was attempting to raise money for the Berkeley Free Clinic along with their sponsorship of the program. Leopold's offered to match every one dollar donation with two dollars of their own, the money to be sent to the Free Clinic in the name of the person donating. Nemerovski said he worked hard producing 18 different commercials for Leopold's to run with the retrospective. He spent five nights producing them. He did it to get the account and because he knew Leopold's did not want to sound like every other record store commercial. Also he believed in prime-time block programming. He thought block programming was better than packages and selling spots, an advertiser got more impressions in the same amount of time so you could sell the time for more. But you had to fool around with promos, and the agencies and other people were against it, and he could see why, it required more time in advance and a lot of extra work, it was a pain in the ass.

A review of the KSAN-KMPX retrospective appeared in the *San Matean* of April 21. It noted that at one point during the program:

> A 1967 tape of Larry Miller, expounding on the reasons why he shouldn't play the Animals at three in the morning, prompted calls from loyal Animal fans, who thought it was all happening live. Miller, married in Golden Gate Park during the Summer of Love, worked for KMPX up and through the famous strike, until he was ostracized from the Donahue family for being a "scab." Another tape featuring morning man Bob Postel, who lived, as well as worked, in the original KMPX at 50 Green Street, even included our South Bay's Chocolate Watchband, singing Dylan's "Baby Blue." The special concluded with a tape of the original Moby Grape live at the Avalon Ballroom, New Year's Eve, 1966-67.

The week after the retrospective, on April 6, Gleason in the *Chronicle* recalled the beginnings of KMPX and compared the station then to what FM rock radio was today. He referred back to the time in March 1967 when he had talked with Donahue at Mills College about looking into KMPX and possibly taking it over, then went on to refer to Donahue's recent KMPX retrospective on KSAN and to mention the fact that airplay on KSAN was now reported in the trade papers of the record business and monitored by music listening agencies. They treated it like they treated top-40 stations. At the time of KMPX, he said, that had not been so.

Gleason's column of April 16 ran with a photograph of Donahue in a KSAN Jive 95 tee shirt standing before a microphone at Pacific High Recording. Dymond clipped the column and the photograph and had copies reprinted for use in station sales promotion.

Gleason said he was prompted to write the column in response to an article in The *Night Times* of April 5 which he felt was not very good. The *Night Times*, a new local entertainment paper, had run a special report on the truth about Bay Area radio with headlines saying that San Francisco AM rockers were vying ruthlessly for teenies and underground stations were spinning hits all the way to the bank. The report referred to KSAN as the *Rolling Stone* of the radio dial, meaning it had a hipper-than-thou attitude, was a top dog, and was as committed to commercialism as to music. The report included a photograph of Street in the KSAN studio on page two. The back page of the issue had a photograph of Donahue and Hamilton in the studio. There were quotes from Donahue, Street, Ponek, and Fong-Torres on what they felt about conflicts between commercialism and the free-form format on KSAN.

The commercialism of FM rock radio was also the subject of one of Wassermen's *Chronicle* columns on April 3, and a prior column of Gleason's on April 10. Gleason in his April 10 column was bothered by the way stations mixed voices. In the olden days, he said, radio commercials were done by voices distinctly different from the voice which read the news or announced the records. The trouble now was that the two were the same and this undermined the credibility of disc jockeys and newsmen. He said he suspected the trouble began when the disc jockeys and newsmen started to read the pitch for waterbeds, or whatever scam was current at the time, and the same voice transmitted the obviously phony pitch as transmitted the real information. It was particularly obnoxious when a newsman followed his newscast, which was at least ostensibly truthful, with a scam commercial. Gleason did not name names in his column, but he later said he meant to refer to McQueen's voicing spots for Undulator Waterbeds with the KSAN news, and they knew it at KSAN. After the column was out they got mad at him.

In McQueen's mind, voicing the Undulator spots was not undermining his or the station's credibility, but lending it to Mike Lavin's business, and because Lavin was a merchant who stood behind his products, McQueen felt that was all right. McQueen slept on a waterbed himself at home and he said so occasionally when he did a commercial with the news. Gleason said he believed KSAN with all its commercials was coopted, you could say it was, but in the beginning with KMPX that had not been so, and the reason was they could not coopt KMPX. They could not because they did not understand it. KMPX was a quirk.

Chapter 15

KSAN
(May-July 1972)

In these few months the individual most important to the station's early period returns as general manager and a staff meeting in his support assumes a significance comparable to the strike of four years previous. This meeting, his response, and a later engineering meeting give evidence of a concern that the staff not lose touch with the audience and parts of themselves that may be important to the station's survival. There is also repeated admonition that they realize what they are doing and not do it blindly.

The General Managership (May-June)

Donahue became general manager of KSAN on an interim basis on Monday, May 15, and he wanted the job, he really wanted it. Prior to that he had been doing his Saturday night show with Hamilton and consulting on programming for KSAN and for KMET. Duff had told him in March he was going to leave. Duncan had come out at one point in March to go with Duff to Los Angeles to fire the general manager there. At that time, Donahue said, Duncan asked him if he remembered when he had asked him a while back about managing KSAN. He had said someday Duff was going to move on with the company and they might want Donahue to manage it. Donahue said he told Duncan, "Come to me then."

In April, when Duff formally announced his plans to leave, Donahue said Duff told him he was going to recommend him for the general manager's job. But he did not hear from Duncan about it. Finally he called Duncan, it went back and forth and he felt Duncan was giving him the run around. In May, on the Monday of the week Duff was going to leave, he was leaving on Friday, May 12, the job had not been settled. Donahue said Duncan told him he did not have the business background. He told Duncan he had a business background more than Duncan thought, but he was beginning to doubt the job would come through and so was Duff. Then on Wednesday, Duncan called and said he thought Donahue had understood he was hiring him for it. Donahue said no, he had not understood. Duncan said well, for a while go in there and manage it as a consultant and after a couple of weeks he would come out and see how things were going and they would firm it up.

Donahue said then he asked Duncan for more time. He told him he needed at least two months or 90 days, maybe he said 90 days, to demonstrate things. They talked for a while. He really wanted the job, the station was his baby. He was still doing consulting for KMET and that Thursday evening he came up from Los Angeles for the going-away party the KSAN staff was having for Duff at the Orphanage restaurant in San Francisco. It was about 5 p.m. when he got there and people were congratulating him on his appointment. The next day, on Friday, he came into the station and watched Duff clear out his desk. That was all the breaking in he got. He had wanted to know how Duff had handled the business and especially how he had dealt with Metromedia, but Duff just cleared out his desk and left.

He got help afterward from Campbell. He knew the company was worried about the business so the first thing he did was pay attention to the sales department, and he gave Campbell a raise, from $210 to $240 a week.

On Monday, June 19, after Donahue had been interim general manager for about a month, Duncan came out for a visit. Donahue said he expected Duncan was coming out to make him permanent. It never occurred to him he would not be made permanent. But when Duncan came, he said what he was going to do was bring Dan Tapson over from KNEW in Oakland. Tapson was sales manager at KNEW. He was bringing Tapson over because they needed a businessman at KSAN. Tapson would be their new general manager.

Donahue said Duncan then asked him if he would consider WNEW-FM in New York, being program director there. They were sitting in the office at the station. He told Duncan if Duncan would do something like average the salaries of the top programming staff at WNEW-FM and add them up, which amounted to a sum of about $85,000, he would consider going. Hamilton was there and she said no New York City, she did not like dog shit. Duncan then asked him to stay on for a while and break in Tapson. Donahue said he told Duncan no, he would just leave the way he came. He started clearing out his desk. By then he was resigned to it. That was the way things went in this business. You had to take it and see the humor in it. He had a pain in his back, he thought it was flu. Hamilton went out and told a few people. Then they went home and he really got sick.

People from the station called him at home that night when they heard and asked what they should do. There were lots of separate calls.

Donahue said he flashed back to the KMPX strike and he did not want any part of it. He told them he did not know, he did not care, he was sick, he could not adivse them, he would not advise them. The next day, Tuesday, the staff had a meeting with Duncan and Tapson at the station. They were furious and they gave Duncan and Tapson hell. They told him about it afterward but Donahue said he never listened to the tape of it. McClay had made a tape and so had Laughlin. He did not want to be prejudiced by what people said about him at the meeting, he did not want to have it on his conscience. Maybe some other time he would listen, maybe a year from now. He got calls after the meeting from O'Hair and from McClay and Bear. Duncan called him the next day from Los Angeles and said in the cold gray light of dawn, he thought he had made a terrible mistake, he thought it was going to take a couple of days to get it worked out, and he asked Donahue if he would go in there for the rest of the 90 days.

Donahue said he told Duncan yes, but he did not remember 90 days. Maybe he had said "give me 90 days" somewhere in the beginning, he was not sure. He could see how Duncan might use that someone had promised him 90 days as a face-saving gesture in the company and he let it go by. He felt it was an issue of fairness. He knew the acting West Coast vice-president for Radio had not said 90 days. Duff had not said it. Duncan had not said it. He knew he did not have a 90-day test but he let it go by. He talked to Duncan about the meeting with the staff. Duncan was concerned about Fong-Torres having been there, that Fong-Torres might write an article for the *Stone*. Fong-Torres had taken notes. Duncan wanted him to call Fong-Torres but Donahue said he told Duncan no. Duncan then called O'Hair at the station and told him the tapes made at the meeting had to be destroyed, because of how it would look to the company.

Donahue said he felt Duncan must have been impressed by the statements made by Campbell and the salesmen at the meeting, they carried weight. For himself he did not want to get involved, and he was so sick he could not. It turned out he had kidney trouble. He knew the staff felt they had done something terrible at the meeting, and he could see they had more to lose now at KSAN then they had at KMPX when they went on strike four years ago. The Duncan meeting was more a revolutionary act for them than the KMPX strike. At KMPX the station was what was revolutionary, the strike was to save it. The meeting this time was their real defiance of authority.

The meeting took place on Tuesday afternoon, June 20, in the Lizard Lounge, one of the larger rooms in the rear of the station. The Lizard

Lounge served also as the office of the music director and the public affairs director. It had been the general manager's office when the format was classical and then was the program director's office for a time. Almost the entire staff assembled for the meeting: disc jockeys, newsmen, salesmen, clerical and production people. They opened the windows to let in air as the day was hot and the seating close and for about two hours engaged in discussion of the following kind:

Street: Why? is the only thing I want to know. Why?

Duncan: This is Dan Tapson sitting here. I asked him to come over [from KNEW] because if he's going to be here he should be in and know everybody and how they think, you know, the best of it, the worst of it, and in-between.

Most of you know that the format's been on four years, four years April. Most of you know that I've backed up the format, since four years ago, even before that, when I met Bob and Tom in the Miramar, and met with Stefan, he had a one-show kind of thing, and convinced Reid Leath that he should go all the way and let Bob and Tom put it together.

What has happened in the four years is that the station has been a fantastic programming success.... It has been getting stronger all the time.... In the four years, the station's been allowed complete intellectual freedom.

For four years, the station has not returned any money on Metromedia's investment, not because it hasn't been successful but because of a series of managers, from Leath through Willis, who have basically been program oriented and not business oriented. They were not able to turn any of the programming successes into revenue returns.

When Willis left, one of the pressures that was laid down was okay, now here's a chance to get a guy who understands business and sales and will be able to add that dimension to the station that makes it complete.

With Tapson, I can go to John Kluge [the president of Metromedia] and say now here's a guy that you know, that represents many years in the company, that's been successful, a man that you can identify with, who is now going to be giving you the facts. And since you can identify with him, now the station is rounded out and just leave it alone.

In my estimation, if I did not ask Dan Tapson to come over to this station, John Kluge would sell it within a year.

I believe we should keep the radio station. I interpret my job as keeping the bloody radio station because I think it's great.

Street: Yeah, but do you want to keep it in working order like it is now or do you want to keep it with all of us here going down, down, down. I mean, we have to work here in order for this station to work, and if we don't work here this station dies, and I think Kluge should start getting hip to that fact.

Duncan: The budget for this station is $750,000. The company has spent $750,000 a year for this station in the past three or four years. That's a lot of money. It hasn't got a nickel back.

Laughlin: You got a radio station they could sell for five times what they paid for it in this market today. The thing was a cheap shit piece of stuff that was picked up by a conglomerate, moved by Kluge up from a half price five years ago, and he's now got something that's the number one property in FM today.

[General agreement voiced]

McClay: You can't ignore the appreciation of the value of the radio station. It's worth five times what he paid for it.

Simmons: Four weeks man, that's not very much, and he's had this place higher than it's been in years in four weeks.

Duncan: I don't think the station would have a chance if I left him in here.

Street: It wouldn't have a chance without him.

We've worked with Tom from way back and we watched Tom grow. He started the whole fucking thing. This was his brain child, and nobody seems to understand that.

McClay: What you're saying, George, doesn't really hold water. It doesn't really make sense, other than John Kluge's whim.

Duncan: I'll tell you what I suggested to Tom...that I would bring over Dan Tapson to work with Tom.

[Laughter]

Simmons: You can't say that to somebody.

McClay: I've heard these stories so many times, George.

Street: Yeah. You can be king but this other king is going to have authority.

Simmons: We're going to bring in Bill Drake and Bobby, you're still music director.

[Bad raps of Tapson, especially by McClay]

Duncan: I've asked Tom to be director of programming for KMET, WNEW, and KSAN. I've asked him to consider all these bloody things. That's evidence of my respect for him. We were sitting here at 8:00 last night and he said let's get together tomorrow and rehash this stuff. We can talk about it again. Obviously he was disappointed and hurt by what I had to do.

There are always people in Los Angeles saying what the hell are those people doing up there?

McClay: We think we're showing them the way.

Duncan: The bomb thing...Foster and Kleiser....Kluge, he looks at ARBs and P and Ls too, Bob. I'm not arguing this point, you're a programming success.

Street: If you're going to bring KNEW over here, why don't you at least let us go over there and use their groovy equipment?

Tapson: My background...Phildadelphia, a salesman six years ago, been in radio six years.

Dusty: When we first started KMPX.

Tapson: New York three and a half years with the rep company. KNEW for two years.

[Meeting is interrupted by a phone call said to be from John Kluge]

Laughlin:	Make it a conference call.
Street:	We should all get on the phone and say burn baby burn.
Tapson:	I've never met Tom Donahue myself.... I have no expertise in this format at all.
McClay:	If they're asking us to accept you, they're asking us to accept what we know to be a mistake.
Simmons:	It's part of our life. It's part of everybody's life here [versus Tapson seeing the station as just one more of Metromedia's stations, just a property, implied by his comments].
Street:	What makes any radio station run...
O'Hair:	The radio station is in this room. The other thing is a fucking license.
Street:	And the thing that makes the people in this room go and want to boogie and go full tilt is a man named Tom Donahue.... The cat brings us energy. We're all a bunch of fucking downers, man, and when Tom comes in we fucking get our shit together, because we know that that cat understands the programming. Willis was one frustration after another.... George, what is the possibility of us buying the station from Metromedia?
Duncan:	I don't want to sell it.
Street:	If it starts dying will you sell it?
Duncan:	The thing about putting Tom in here came about April. Willis up and quit. Tom was better than putting someone in here who didn't know anything about it.

[The question of Tom's "contract" is raised, an interim period of 90 days is referred to.]

Duncan:	Willis said 90 days. Here's the decision. Dan is coming over. I'd like Tom to stay. I can see that Tom's hurt.... I'd like Tom to stay forever.

Ponek: Yesterday everybody here was as high as they've been in four years. All of us will care about what we are doing here. And today man nobody sees any reasons to care because it doesn't make any difference. You have no conception as to what it is that makes this station great. I don't think anybody on your level can conceive that the difference in there is that everybody cares to make this their ideal radio station. That's why it gets ratings.

McQueen: It's a whole lot more than a job because everyone of us has given up all of his energy. I used to do a lot of other things in addition to my job. I'm a good painter and photographer and I used to do a lot of that. But this job takes all of my energy and I've given up things that in a long-range sense are more important to me.

Someone: Everybody here has put up with Metromedia for four years.

McClay: A critical question: What do you think if this radio station, meaning the air staff, most of it, if not all of it, and the sales staff, most of it, if not all of it, walked out on the street, quit, and took themselves as a package to another radio station, with the ARB book.... What do you think would happen to this radio station?

Duncan: Well, this station would cease to exist.... This isn't KMPX again, we're going to let you work.

McClay: KMPX and what Leon Crosby did to Tom Donahue, that symbolic act, we're seeing this repeated now.

Duncan: It's really not the same thing.

Ponek: Yeah, it is, George....

You sell the whole thing out when you go back on an agreement with the man who created this thing. And you put somebody in with no understanding of what it is that has made us great.

McClay: This amounts to a slap in the face of our life-style.

Duncan: Willis knew absolutely nothing about this radio station when he walked in the door. He had never heard the goodamn music.

Gossett: That's true, and it took us four years to get him straightened out.

Duncan: But he added something.

Street: He added a lot of trouble.

Duncan: Whatever he did he couldn't have been bad because he didn't get in the way of your continuing to grow and flourish.

McClay: He had an intellectual sympathy... before he ever came here.

Ponek: Why take Tom away? [said very plaintively] He's the man who really knows how the thing runs.

Simmons: Even Donna Campbell wants Tom Donahue.

McClay: Metromedia.... The New York office is shot through with imcompetence.... Metromedia's getting shot through from the top.

Simmons: [re Tom possibly going to New York] He doesn't want to leave his kids. You know man, this is his family.

McQueen: What you're up against here, George, is that you've got, we think of ourselves as, and I think we are probably the best and most professional, innovative, and creative radio staff in the whole country, and Donahue is just about the only person we can think of who is not even our peer but our superior in ability.

Laughlin: Isn't there a shake-down going on right now in the whole of Metromedia?

McClay: I would suggest that a very wise decision for you to make at this point in time is to back down on your decision and at least give Donahnue a chance to do what he said he was going to do.... Or Metromedia and us are all going to come to a parting of the ways. It's like a divorce.

Tapson: As professional radio people, why should you suddenly stop?

McClay: We're not going to stop. We'll go to another radio station.

Street: We've been shafted too many times and this is the final straw.

[Comparisons with Metromedia's treatment of KNEW made]

Someone: We've been shat on and shat on and shat on and this is the worst.

Dunlop: I loved Willis, but Willis didn't have a good business sense. I loved Whitney, too, but Whitney would put business on the air when chances of collecting might have been scarce. That's bad business. Donahue has shown in three weeks that he's got an insight into this, and Willis didn't show it in three years.... Donahue is so right, right now, for it.

Simmons: The record companies respect him.

McClay: It ain't going to be here if he leaves.

Street: He has the energy that makes me want to do my very best man, and I haven't wanted to do my very best for two years.

Laughlin: You got to remember we're mostly ex-professionals here. Donahue originally gathered to himself people who had become so fed up with this sort of corporate bullshit within the broadcast industry that they had dropped out of it but had great reps in the markets they were on in previously.

Someone: What's wrong with Donahue? Why?

Duncan: He is an unknown quantity as a businessman, and there are businessmen running this corporation.

Street: Was Willis known as a good businessman?

Gossett: After Tom's gone out and used his reputation to get in a lot of back money and everything. I think it was a blow to his pride because he's wanted this station ever since he's been here.

Duncan: Tom feels he was taken out of a position that I never felt he had.

Dunlop: Within the first three weeks he's called in 6% of the outstanding bills, hasn't he?

Ponek: Tom's not going to suffer the indignities of being given a throwaway job... .

McQueen: Exactly what authority would Tom have?

Duncan: In the programming sense, almost complete authority. Is that fair?

Laughlin: What's the "almost"?

McClay: Is this so important to you that you would destroy the radio station to do it?

Duncan: Well I don't see it as destroying the radio station... . I can't see how the station would be ruined unless you ruin it. Nobody's going to ruin it for you.

Laughlin: John Kluge and everybody back East in this corporation considers that we are a bunch of lunatics who have taken over the fucking asylum. Now, right now, what we are seeing is a changing of the guard because they're not going to let the boss fucking loony run the bin.

Street: Right.

[General applause]

Laughlin: John Kluge doesn't trust anybody whose hair is longer than his teeth.

George, it isn't as if we could call a strike man, because we're all aware, the ones who went through the KMPX strike, are well aware that there have been four years of strikes since then and nobody gets it that way... . Metromedia doesn't lose any strikes.

Street: All of us have other things that we could do man... . We don't need the station to exist. It needs us.

Laughlin: We all know that you've got the property and you could bring in religious programming, you could do anything you want with it. The question is do you want to do that?

McClay: Individually each of us is replaceable, but collectively we are not, and collectively we are worth something.

Laughlin:	And collectively we're pissed off.
Ponek:	You're going to end up firing us one by one, because some crony in New York is, you know, maybe a big contributor to the Nixon administration. You might as well do it now man.
Boucher:	Why are we all hung up in the past, why don't we look in the future?
Someone:	We are.
Laughlin:	We are projecting three or four months of good weather between now and the first rain.

Substitutes filled in on the air while the staff met with Duncan and Tapson Tuesday afternoon and there were periodic mentions that the regular staff was in meetings and was having trouble. At 10 p.m. Tuesday night, Street went on in her usual spot. She opened her show with "Slippin' into Darkness" recorded by War and introduced it by saying, "I want to dedicate this record to Tom Donahue and a lot of other people who are slippin' into darkness." O'Hair the next morning played the same song at the start of his show at 7 a.m. and said it seemed to be the company theme song at KSAN: "You been slippin' into darkness, pretty soon you gonna pay."

O'Hair said after the meeting late Tuesday afternoon, everyone ran for the phones. They called Donahue, they called Wasserman. The call to Wasserman ended up as a conference call from the front production booth. Duncan and Tapson left and called back up from downstairs, from the Metro Radio Sales office on the third floor which had a private line into Campbell's office. O'Hair said they asked him to keep it quiet, cut publicity. A little later, Duncan called from the airport about the possibility of a *Rolling Stone* article and asked him to keep Fong-Torres from printing one. The next day, Duncan called from Los Angeles. It had to do with justification for changing his decision. O'Hair said he told Duncan the staff would walk out for him if he was threatened for his job by Metromedia as a result of deciding to keep Donahue on. They would support him like they had Tom. Duncan said he really appreciated that. O'Hair, thinking back on it, said he felt everybody got stroked. That day, June 20, was his thirtieth birthday and he had been stoned on acid since the morning. Duncan at the time was 41. Donahue was 44.

Duncan said he changed his decision to replace Donahue as general manager because of his conscience, and not as a result of intimidation by the KSAN staff or the press. He realized his original decision had been based on a wrong set of criteria. It mattered to him that the department heads at the station were for keeping Donahue on, that was more important than the staff wanting him. After the meeting on Tuesday he was steaming. He went down to Los Angeles to see the West Coast legal counsel for Metromedia and his hand was sore. He had pinched a nerve while squeezing it into a fist during the meeting. He felt the staff was putting great pressure on him and there was a threat of the press, which was negative pressure. Having it out in the press would take it outside the family and the more they leaked, the more handicapped he would become. What he wanted most was to be able to be objective. When he left the station he knew he could not make his assessment in response to public pressure, he would be a fool in the eyes of the whole United States if he did that. He had to do things which would enable him to be objective and make the right decision. The staff in their meeting almost ruined it. Down in Los Angeles he decided, based on his judgment, to reinstate Donahue, and he felt his decision to do that was not only popular but right, and good for the company. His decision to move Tapson in had been premature and it may have been wrong. He would wait a few months and see how Donahue worked out. He liked Donahue. He had put him in when Duff left.

Not long before Duff left in March, Duncan had come out and suggested to Donahue that he work on the programming at KMET. He said it was while he was having a beer with Donahue and Hamilton. He thought it would be perfect having Donahue on both KMET and KSAN. Originally it was on a consultant basis. Then Duff decided to go and he put in Donahue as general manager at KSAN. He was not sure where Donahue's 90-day term came from. He himself had not said it, it was said by someone else in connection. He thought it had West Coast origins. He had not remembered when he brought Tapson over in June that it had been said. He went out there and he was hit with the accusation he had gone back on his word, that he had promised Donahue 90 days. That would put off the determination of whether Donahue would be permanent until late August or early September.

Donahue said he heard later that Tapson had been hired a week after he took over in May to replace him as general manager. Duncan had known in advance he was coming out to introduce Tapson to the staff. Everybody in the company knew but him.

Fong-Torres said the story he might have written for the *Stone* got dissolved, it got dissipated in the reversal of events. Metromedia did not want a story, the *Stone* did not want a story. He thought at the time he might do a quick, short piece but that was not possible, it was outdated too fast. The other alternative would have been a longer feature but the *Stone* did not want one for the next issue. Then his brother died, he was shot by a gang in Chinatown, but he would not blame it on that. Afterward it seemed to be dead. He knew Duncan was upset but he had intentions of doing a story anyway. He had talked with Donahue and had a quote he wanted to use. He had notes from the meeting, he had been there.

Wasserman of the *Chronicle* had not been at the meeting but he had been called by the staff and he added a note at the end of his column of June 21:

> Tom Donahue, the man who "invented" the underground FM rock radio format at KMPX in the late '60's, has been fired by Metromedia as interim general manager of KSAN.
>
> Dan Tapson, sales manager of Metromedia's local KNEW, a MOR station specializing in oldies but goodies, will take over the top KSAN job. The KSAN staff, meanwhile, is not particularly happy and had a meeting scheduled for last night to discuss the situation. More on Friday.

On Friday, Wasserman reported the firing and rehiring of Donahue in detail.

> Well, the great KSAN rock crisis is over, almost before it began. Dan Tapson is still sales manager at KNEW, Tom Donahue is still acting general manager at KSAN, and all is right with the world.
>
> The word that Donahue, noted Big Daddy of local rock radio and innovator of the free-form FM format, had been canned came as a regular lightening bolt on Tuesday afternoon, inciting a mad humming of telephone wires and an angry staff meeting at KSAN. As someone said at the time, "We're trying to figure out whether to walk out, sell out or do something in between."
>
> The entire mess may be traced to the May departure of the venerated general manager Willis Duff, youthful prodigy of the wireless. Donahue was hired at the last minute as interim GM. He was to run things for three months, then be considered, on the basis of his record, for a permanent contract.

Bear said he called Donahue after the meeting on Tuesday. He had not said anything at the meeting because of his past with Donahue and because he felt the people there were being vicious. When he called Donahue, he told him he would not believe what just happened. He estimated about half of the talk at the meeting was in support of

Donahue, about a quarter was against Metromedia, and a quarter was directly against Tapson. Most of what was said was repetitious. In the course of two hours everyone but himself stood up and said something.

McClay said he felt the meeting was like the KMPX strike and that Boucher kept saying "Come on, we have to see both sides," but that was not it, they had to come down all together. The day after the meeting, on Wednesday, when McClay came into the station just before 2 p.m. to do his show, he said all he wanted was to do his show. While he was on the air, Donahue called in the news that Duncan had changed his mind and decided to keep him on. Different people around said, "We won!" McClay said he felt it was a more important victory than the KMPX strike, this was a real victory. McQueen announced it on the 5:30 p.m. news and said now they were going to try and make it the best radio station in the country. Street said she felt nothing could have gotten the staff together but Donahnue's firing. That was what got them together before, when they went on strike from KMPX. Street said at the meeting Campbell threatened to quit if Donahue went, and when Campbell said that, it made her feel like crying.

Campbell said she told Duncan before the meeting that she wanted Donahue at the station. He announced it at the meeting. She did not say anything then. After the meeting she was on the phone to Duncan. She felt Duncan and Metromedia listened to her. Over the years she had gotten a reputation, they valued her advice at the upper levels.

Boucher said he felt the corporation was irrational and that Kluge the president was crazy to put Tapson in, the station was making money. He thought, and so did McQueen and Kibler, that the station would be up for sale soon. The lease for their floor in the building ran out within the year. Kibler told them the building was owned by Jesuit priests.

O'Hair said he wanted so much for Donahue to be general manager after Duff, just to see what would happen. He wrote in his monthly report for May: "Donahue has been named Interim General Manager. If queried how it felt to work with Donahue after Duff, I would have to reply, 'It's like having Johnny Unitas called out of the game to be replaced by Joe Namath'." Campbell was also grateful for Donahue's coming on. She felt he took her into his confidence. She wrote in her monthly report for May: "Tom, I appreciate the communication between your office and mine. It makes my job easier and I can better serve you."

Duncan said he felt when Duff first came to the station in April 1969, it was anarchy except for Donahue, and Donahue did not have the authority or desire to run it. Duff came with an ability and knowledge to function within a corporation. He also had a willingness to become friends with the station, with the people there, and to become assimilated into their culture. But there came a point, Duncan thought, where you could not be friends and assert your authority, you could not say "that stinks" to a guy you grew up with. Duff ended up with trouble on that account. But when he left, Duncan felt, the station was a cohesive, successful unit. When Donahue came back, he inherited from Duff a radio station that was professional and had a personality. What Duff had inherited from Paulsen was a bunch of people with an intellectual commitment. Duff had to live with that. Duncan said Duff was on the phone to him a lot. Duff would get mad at him on the phone and he would have to tell him what to do with the corporation.

The staff tended to deprecate Duff after he left. It was not that they had not liked him, but he did not benefit from comparison with Donahue. Hester said he liked Duff but he felt Duff was a businessman while Donahue was a righteous cat. Gossett felt Duff was a businessman and a scholar and an easy person to propose things to. He was inclined to do things first and worry about what happened after. An example was Winterland one time when it got messed up. Donahue thought before about consequences, he was willing to gamble but he thought before.

Donahue said he felt what he did was act on instinct, maybe he took calculated risks, but he was not a rational calculating sort of person. He felt Duff had wanted acceptance from people at the station when he came and over the time he got it. But he did not accomplish some things, he had a tendency to accommodate people too much. He took the stance of, "I'm just a country boy here to make friends with you." He tried too hard to make friends, to become one of the boys, and that left him helpless in the end. Donahue thought maybe he would speak differently about Duff if he had not followed him in the job. But he had and he felt when he started he got all of Duff's dirty laundry, he got everything Duff let go or never cared about.

Donahue said when he turned the station over to Duff in April 1969, he saw that Duff intended to make the station into a well-running business operation, and he felt Duff left in May thinking he had. But he had not. He had gotten coopted by it. He became caught in the craziness and it all went on just the same as it always had, and still did. But the Duff who left was a very different man from the one who first came. Donahue said he and Hamilton told Duff in the beginning that he could be a secret hero of the revolution. Hamilton felt Duff believed it. When he left they gave him a small pouch of coke for having been a secret hero. Duff said he knew the secret hero line was a joke but he did feel some of his basic values had changed. He had always been conventional, conformist to whatever culture he was in. In his last two years at the station, he felt he became more eccentric. He looked different by the time he left, his hair was longer and he had lost weight. He had split with his wife. In the beginning, his salary was $25,000. It was $33,000 by the time he left. He did get involved with the staff, but he felt he was never really personally close with any of them, although maybe with O'Hair, and with Nisker more than anyone.

Donahue was not very comfortable when he started as general manager in May. He thought back to when he had been program director before Duff came. Then he could say to the staff, "Hey, do what you are supposed to because I don't want to have to tell you to." As general manager, he had to lay down the law more, he was expected to be the seat of authority. But there was a question as to how much he could really do, and he knew the station would not be a democracy, it had to be a patriarchy. That was what Wavy Gravy had at the Hog Farm. He knew he had a big daddy image already. He had got it in top-40 radio. He felt it was useful, he could hide behind it and protect his privacy. But it surprised him sometimes because he would forget. Then he would walk into the record library and everything would go quiet and he would realize it made a difference. He felt he had a problem now about whether he had to be there at the station every day. He wanted to work out of his house more and when the staff wanted him they could call him, but they seemed to need someone in that office.

Conditions (May-June)

On Monday, May 15, the day Donahue started as general manager, tickets went on sale in San Francisco for four Rolling Stones concerts scheduled for Winterland June 6 and 8. These would be the second stop in the Stones' upcoming tour of 31 North American cities. The last time the Stones appeared in the Bay Area was at Altamont in December 1969. The Winterland concerts were sold out the afternoon of the day they went on sale.

On Monday, May 15, Wasserman noted in his *Chronicle* column that residents of Berkeley were celebrating the third anniversay of People's Park.

On Friday, May 12, the day Donahue came into watch Duff clear out his desk, there was an anitwar demonstration scheduled for noon downtown in Union Square. Waterbed Experience called the station to tell Nisker he did not have to run their ad with his program if he felt it would be offensive along with his coverage of antiwar activities. Sanders took the message and asked Gardner what he thought she should do about logging it. She was not sure how to write up an advertiser who had agreed to sponsor a program and then offered not to have his name aired. Gardner told her he thought it was nice of Waterbed Experience to offer but it could be more trouble than it was worth.

On Wednesday, May 24, Donahue went on the air with a two-hour talk show on which he invited listeners to call in and tell him what they thought of the station. The show opened at 10 p.m. with Donahue making the following statement:

Donahue: I'm hoping during the course of the evening that I'll be able to get some positive ideas from you and that you'll be able to get some rational explanation from me. I've been with the station since we started this kind of programming which is sometimes called progressive, which isn't really a very good title. Other people call it underground which it obviously isn't. I used to call it free form because that didn't mean anything. My new choice is alternative because that doesn't mean anything either. If you'd like to talk about it, we'll give you some telephone numbers to call and I want to remind you that we are going to be on a delay, so you bad talkers will not be able to expose yourselves and get you or me in a lot of trouble in the next two hours... .

My association with the station started in May of 1968 when we started this kind of format. I was at the time the program director and was on the air with a record program from (when was I on the air?) 6 to 10, and I left the station in about April or May of 1969, and since that time I have just been doing Saturday night shows. Recently I was appointed general manager of the radio station on an interim basis. Now let's face it, everybody's on an interim basis. But I think the message there was the company wants to find out if I can cut it as the manager, and if I can cut it I'll probably have the gig. If I can't cut it somebody else will have the gig.

Of course, I look upon it also as I want to know if they can cut it with me. So it's sort of a romance that may or may not lead up to marriage or the alternatives of cohabitation. I'm going to give you the telephone number, then I'm going to play another record, then if you'd like to call me and talk about the radio station we'll talk about it, if you'd like to not call we'll play records. The telephone numbers are 986-6244 and 986-2825, burn those into your mind, put them next to the fire department number, 986-6244 and 986-2825.

It's your radio station, the Federal Communications Commission is very clear about that. Licenses to broadcast are handed out to people on the basis of their functioning in the interests, convenience, and necessity of the community, and that's what we want to do and that's what we want to know about from you, you know, what you think are interests, needs, conveniences, necessities.

Most of the listeners who called during the evening complained about the station's commercials. News was the second most frequently mentioned subject, it got some complaints but mostly praise. There were several questions about the termination of the gay and women's shows, several about the kind of music played on the station, and several about the money made by people who worked there. Donahue tried to assure a few of the callers that whatever they thought, KSAN was not the same as a top-40 station, they should turn their dials and listen. He said none of the people who worked at the station were getting rich off it. He described the professional backgrounds of some of them and the ways they were trying to obtain more control over commercial copy. He said he felt offensive commercials were bad, not only because they irritated the audience, but because they failed to sell the advertiser's product. He told a few callers no matter how many commercials they thought they heard, the station still had a policy of eight spots per hour maximum and suggested maybe what they were hearing, which made it seem like more, was commercials in time periods which had not had many before. Several callers asked why the station did not turn down Bank of America ads in opposition to the war in Vietnam. Donahue said Bank of America ads were not on the air at present, but the station had run them in the past and still might, and maybe his philosophy was too pragmatic for some, but he did not think turning down Bank of America would stop the war.

One caller asked about Metromedia and what their attitude toward the station was, did they have feelings about what was going on there or was it strictly a money-making proposition? Donahue answered:

Donahue: Well I think that they've got a real interest in KSAN from the point of view of an operation that they've allowed to function in the manner in which it has. We are, I think, an anomaly among radio stations, particularly among commercial radio stations. Metromedia is a company that has been willing to do controversial things over the years. They were one of the first radio station chains to go into talk radio, and talk radio brings about complications that bring you constantly in trouble with the Federal Communications Commission. This radio station has from time to time had difficulties with various government agencies, and they have not as yet made any great amount of money out of the radio station. This is the fourth year they have owned it and it made money one year and that was on a percentage smaller than could have been made if the same amount of money had been put in the bank.

Another caller asked about the death of underground radio, why was it dead? Donahue replied:

Donahue: Because life is a series of moments man, that you dig while they're happening and don't expect to constantly repeat them. There were many elements of radio, let's say, in the early days of KMPX, that were glorious for the moment. For instance in the early days of that radio station we had practically no commercials, I suppose there was a great deal more music. In that situation as the station grew more successful we were able to get more commercials on and started to pay a lot of people who were working for little or no money. So there are many times I think, ah, we really had an incredible amount of fun and that was great, but I also think about the fact that we are now able to do many things that we weren't able to do in those days. If you remember we never had any news on the air in the early days of KMPX. When we first came here we had an ABC network news service. Now instead of that we have a news department of our own that really attempts to present news that has credibility and I think that in this country today news with credibility is truly alternative news. So the fact that we have gained increased economic success makes that kind of thing possible for us. That is, it's a completely constantly changing thing.

Caller: Would you say that the people that were doing underground radio back then were doing it primarily for the people and were not interested in the money?

Donahnue: No, I wouldn't say that.

In answering another caller, Donahue mentioned several new features that would soon be added to the station's programming. Voco would be starting a sports show on which he would interview athletes. McClay would be doing a show about movies and would be interviewing movie people. Someone else would do a financial news feature, as a good part of the counterculture was in business and they might as well know about it. A psychiatrist might come on to talk about paranoia.

Donahue, off the air in May, shortly after he became general manager, introduced a change in the station's music record-keeping. He asked Simmons and Cole to make up lists each week of the 30 most-played records of the previous week. O'Hair said in his May report the lists would not be used to urge the air staff to play certain albums or cuts more than others, but rather to make their reports to the trade magazines more accurate and to give the music director and record librarian a way of knowing what was being played when they were not there. The lists would also be given to the salesmen so they would have something to show to record companies. O'Hair said in addition they were developing a system to find out what the record stores were selling and what the audience really liked. This was in order to keep the staff from letting their musical tastes become too parochial and to see how effective the station was in retail sales.

Cole said he could see a point in compiling the airplay lists even though it was more paperwork. Duff had wanted him and Simmons to keep the salesmen up on what the station was playing but he had not had them keep lists. Simmons said she had done something like keeping the lists before, but it was more informal. She had been calling in airplay reports to *Record World* and *Billboard* almost since she began as record librarian. But she used to look at the shelf and just pick out new records she thought they were playing. McClay said that was what Donahue did, wherever he went he always created paperwork. Donahue said the purpose of the airplay lists was accounting, to get the staff to see patterns and realize what they were doing, what they were actually playing. He wanted them to be conscious of reasons for playing things and not do it blindly. He wanted them to get out in record studios and clubs and places where they would be exposed to their audience more. He had to remind them that they were older than their audience and they did not automatically know what their audience would like. He had to keep them open or they would lose touch, and then it would die.

Bob Dylan turned 30 in May and Voco wrote liner notes for a new record album he had produced, his first as an independent producer. It was due out on Blue Thumb early in September. He was calling it "Lights Out: San Francisco," which was what he had called the last hour of his 6-midnight show on KMPX Saturday nights and the whole of the show he did on KSAN midnight-6 a.m. Sunday mornings. He said in the notes he wrote for the album he was grateful to Donahue and to the chick engineers, the ladies, of KMPX, and in particular to Street. Street wrote some comments for the album and had her picture on one of the inside jacket covers.

Street was married in May and the *Chronicle* of May 16 carried an announcement of her wedding on the society page in an article titled, "His Bride, the Disc Jockey."

Kibler wrote in his May monthly report that bids had been submitted for construction of an outer door and an enclosed reception area in the entrance to the station offices on the fourth floor of 211 Sutter Street. Donahue said he wanted the entrance to be protected so people would not wander in and distract the staff from work. Someone suggested the glass in the entrance be bullet proof.

Skinner said in his sales report, May was a damn fine month, the best five-week month in the station's history. The gross was up 7% over May of the previous year. The gross in June was up 14%, and in July up 40% from the same months the previous year.

The April-May ARB came out in June and Skinner said it was the finest ARB the station ever had. They showed up as number one in the market for adults 18-34, Monday through Friday, 7 p.m.-midnight. Ratings for the morning show were double those in the previous book. The best of the ARB results were printed up for use in presentations by the station's salesmen. In prime time, between 7 p.m. and midnight on weekdays, the station had 12,400 adults ages 18-34, average quarter hour listeners. Of these, 8,800 were men and 3,600 women. In the afternoons, 3 p.m.-7 p.m., there were 18,600 adults in the age group, 11,100 of them men and 7,500 women. In housewife time, 10 a.m.-3 p.m., the total was 18,000, of whom 11,000 were men and 6,900 women. In the mornings, between 6 a.m. and 10 a.m., the total was 16,500 with 10,500 of them men and 6,000 women.

Kibler felt in June that he would leave the station soon, maybe in August. He had come on as chief engineer a year ago August. He said his reasons for leaving were he had not had a vacation for a long time. The station had no plans or arrangements for substitutes or relief engineers to fill in for him. He was separating from his wife, their marriage could not stand 50 to 60-hour work weeks. He felt that happened to a lot of people at the station. Donahue had replaced Duff and he did not think things were going to get better for him. Duff at least could understand the engineering problems, he had a first-class license. Donahue did not. Kibler felt with Donahue there would be more problems, less understanding, and more pressure, and Metromedia was not spending money. He had tried at the start of the year to get money for equipment to improve the commercial production studio but Metromedia had said no, no capital outlays for the station. O'Hair wrote in his report for June that

Kibler had decided to buy a farm and would be leaving soon. They were looking for a replacement, preferably someone with three arms and a good sense of humor.

Pigg was fired from KSFX late in June. He said he had expected to be fired but he thought it would come two months later. KSFX had adopted a tight playlist policy and cut out disc jockey raps and he ended up reading comics to keep from being bored between reading the formatted raps. When they fired him, they told him it was because they did not think he would be happy there. The day he was fired he came over to KSAN. He did not ask for a job, he just hung around. Donahue was in and he said to him, "Well, Pigg, we can put you on weekends if you ask me for a job." After that he filled in for McClay one afternoon and for Fong-Torres after Fong-Torres' brother was shot.

Coming back to the station, Pigg had the feeling he had gotten more conservative in the time he had been away and that the people at KSAN were somehow still hip, still into psychedelics and living according to a hip philosophy. He knew probably that was not so but he felt it was and he still respected the station. He did not have to respect anyone there to respect it, and what it was, the tradition. If they fucked with it too much, maybe he would not respect it, but as it was now he still did. It had made a big difference to him, people like Bear, back at the time of KMPX saying they had to respect their listeners. In top-40 radio, disc jockeys did not respect their listeners. Pigg felt in coming back Bear was one of the people he really loved. He had never told Bear that and he could not stand to be around Bear long, but still he felt it.

The Staff (July)

Dunlop said he felt with all the changes the station had been through, the salesmen, the three of them, himself, Rick, and Jeff, had remained pure. He knew Donahue was not fond of them. Donahue did not get along with salesmen, he thought all they wanted was to make money. But he, Rick, and Jeff had started as radio groupies, they had really loved it in the beginning. Now it was harder to like but he felt they still wanted the same thing, the feeling of being a team, the excitement about what everyone was doing, that they were doing something you could be proud of.

Dunlop thought Donahue did not know he had started out in show business. He had been a child actor and once was an understudy for the boy lead in Lassie. He cared about things other than being a salesman. He felt entertainment was important, making people just relax and forget their troubles. Otherwise he might as well work for an advertising agency. He felt when Harris was sales manager, the salesmen had been more of a team at the station, they were more equals. With Skinner, it was more divided. Skinner put in more controls, he expected daily reports from each of the salesmen. Skinner had been there almost a year and it felt like everyone was in it for himself now and riding on the success of it. Dunlop said he felt that was what he was doing, riding on the success of it, he liked to be on a winning team, and he was possessive about his accounts. Skinner told him it was not professional to be possessive about his accounts, it was childish. The thing to be concerned about was the station and his job with the station. But he wanted to be enthusiastic like he used to be. There was a time when how good he was doing with the station was all he could talk about with his friends. It was hard on his marriage, his wife complained.

Campbell felt over the last few years she had mellowed. She had become more tolerant of the staff and they had become more accepting of her. She thought most of them saw her as a Metromedia person and she was, but she wanted them to like her for what she was.

O'Hair said he was in it for the money and to build a power base. He felt he had a military psychology and that he had to, to obtain the respect of the staff. Sometimes he was in a position where he knew something he could not tell he knew as the basis for a decision. He had to have the respect of the staff so they would accept the decision, even if they did not like it and could not know why he was doing it. He came into the station every day and stayed in the evenings, and he came in on weekends. He would not take the three-week vacation coming to him, he expected it would take too long to get back into things when he got back, and someone else might have his job then. He had to be careful.

O'Hair felt since he had come a year ago in August, his commitment to the station had changed. At first he was out for his own survival, but then it got so he felt he could create or use issues with a knowledge of their consequence for specific ends. He used favors that way. If he told Street she could do something and it was a favor, he made sure she knew it so the next time he had to tell her to do something, she would know he had already done something for her. He thought of sending a Balls Award to Duncan for reinstating Donahue after the June 20 meeting. The meeting showed Duncan had no balls but also that he did. O'Hair said he was glad Donahue was there after Duff. He could respect Donahue. Donahue was a freak, an odd one, a triple gemini, like him.

Street said she felt one thing Donahue knew was he knew what worked. Duff did not. Duff had crazy ideas for promotions that would never work. Once he wanted the staff to pose nude in a pool for a picture. They told him he was crazy. He said they could all stay underneath and just have their heads appear in the picture. Now there were some people at that station Street said she would not want to be naked in a pool with. She felt in some ways Donanue was still in the old days, he thought they would do the work for love, do commercials on their overtime and things like that. He did not realize they were in it now for the money. Street felt one thing she had learned was you had to be cynical in this business, you had to be cynical in order to survive. You were going to be fucked in the ass and you had to take the pleasure you could get from it. That was better than just getting fucked and being realistic about it. You had to turn the joke. If you went to lunch with one of the salesmen to get a client, you had to think it was on them. You had to believe you were a free-wheeling spirit even if that was not really the way it was.

There were high points. Street felt she had been able to meet her heroes, B. B. King and other blues people, but radio was a business and a short-term thing. An AM disc jockey's life with a station was an average of six months. Street said she had been thinking of leaving radio, leaving the station, for the last seven months, maybe a year, that was why she took time off in the fall. About a year ago, she started getting depressed with her shows and feeling she was going no place. She saw radio now as a means to another end, maybe getting into record production, she might be out of it in a year.

When she first started doing her shift on a regular basis, back in December 1969, Street felt it had been total enthusiasm, she had wanted to see if she could do it. She was gung ho for about a year. She listened during the day and prepared an hour before each show. Then she started getting less prepared and coming in late, sometimes drunk, and loose. It was depressing. Now, compared to the beginning, she felt she was not really working. People said she was great but she did not think she was so good. It was fake now. She knew what was behind what people saw and it really was not there anymore. Some promotion men said to her, why drop out now when she had such a reputation, when she was such a heavy jock? But she felt being a jock was too limiting, especially if you really wanted to change things, and there was the competition. She had been afraid for about a year that another chick would come along and be better than her. The fear was not that she would lose her job, but that somebody better would be on the air, it could be on any station. Then she would have to step down. The reputation had gotten too big for her. She wanted to step down now, to get off the pedestal before she got kicked off.

The fear first came when her ratings got better than anyone else's, when they hit the nines. They had gone up since then, with one drop, but she was still afraid. It had become a standard thing on FM rock stations to have a woman on the air. Mary Turner was on in Los Angeles on KMET. Alison Steele was on WNEW-FM in New York. There were others around the country. Someone said there were more than 20. But so far as Street knew, she was the first.

Hamilton worried about Street, that Street would burn herself out physically before she had used her mind. She felt Street had ego troubles and that often she was two-faced, it showed when she was drunk. She had troubles about being both a star and a groupie. Donahue said a disc jockey could not be both. But in a way they all were. Donahue himself told McQueen one day he was one of McQueen's fans, and he meant it, and he had great affection for Street. He thought some people who got into radio got used to the system and could manipulate it. Others like Street turned out to be good but never reconciled with it.

Donahue said he knew sometimes Street was honest with people and admitted the problems she had with her reputation and the problems she had with men. It was hard when you could not tell whether they were just kissing your ass for something. It was like what happened with Janis Joplin. He did not personally like Street's music but he knew she had an audience, he thought they listened for her personality. But a problem was she had her one thing and she really did not want to do anything else. She might switch shows with Gossett and go on from 6 p.m.-10 p.m., but then she would do Gossett's show like Gossett would, or like Street thought Gossett would, not as if she were Dusty Street doing that show. On her own show, Donahue felt, she would go on for days playing only what she loved if she could get away with it, and people would listen, not just for the music but because it was Street that was playing it.

It bothered Donahue that Street and some of the others at the station did not have a desire to expand their repertoires and become radio professionals. He could, or he liked to think he could, do a country and western show or a classical show or any kind if he had to. It was part of having a professional repertoire. But many of the staff at KSAN had not gotten into this for that. They got into it more as a personal thing, they took the star trip seriously and thought of themselves as minor celebrities. Donahue said it was not that he did not have some of that himself, he had gotten into radio through top-40 and there was a certain amount of stardom in top-40, but it was mainly a profession. It was not so much a personal reflection of the people in it, at least they did not take it that way.

Bear felt he had been influenced by Street to get into the blues and he was grateful to her for that, but he felt about her that she had always been better, stronger, then the men and women around her and you did not develop compassion for people that way. There were times he would come into the studio to do his show just after hers and it seemed to him in playing her music she was crying her heart out. But he could not really talk to her about it. He had put in a request for her shift, if and when ever she left.

Bear said he heard McClay on the air one afternoon a few weeks back playing a set of Latin love songs. It must have been after McClay had a fight with his old lady and he was trying to reach her, he knew that kind of music was what she liked. It was a beautiful set, very sentimental, not what McClay usually played. Usually McClay played what was going on around him. Donahue said McClay had garbage taste, street taste, like his own. But they all sometimes played their music to express very personally how they felt and they each had their styles. Bear said he thought Donahue's style had remained basically top-40 and that Donahue was speaking to an audience that was not there anymore, but it was hard to tell. Ponek, he thought, was still basically classical and country in what he liked, but Ponek had learned from the rest of them.

Ponek each day felt a sense of relief when McClay came into the studio to do his show. McClay had the 2 p.m.-6 p.m. shift just after Ponek's and Ponek would announce when he saw him that the daily miracle had occurred, McClay had arrived in time to do his show. Both Ponek and Bear started music and interview programs on KEMO-TV in July. Bear had his first show on July 17, Ponek had his first on July 24. They had each gone over separately and talked with Crosby beforehand. Crosby said there were others from KMPX who had stopped in and asked about work since he started the television station in February. Some of them apologized for what had happened at KMPX. Ponek told Donahue he was not sure about KEMO-TV but he was giving it a try. Donahue nodded and later said he thought it might help KEMO-TV but he did not think it would help KSAN, his disc jockeys did not look good on the air.

In July, about when he started with his television show, Ponek felt he was finally coming out of it. He felt he was worth something now and he had not felt that way for some time. He had stayed apart from the station after he resigned as program director. He would come in spaced out and do his show as a kind of fantasy trip and was generally low for about a year. Returning to being just a jock after having been program director, he had to live with his failure, he had to recognize his mistakes and work his way out of it. He thought if Duff had been worth his salt he would have fired him during that period, his invalid period, but Duff kept him on and the staff tolerated him. Then about a year and a half ago, he started to put his life together again. He joined Delancey Street and got involved in their group therapy sessions, which were like Synanon games, like the staff meetings they had at KSAN while he was program director. But the staff meetings, he felt, were more vicious and not necessarily therapeutic.

From Delancey Street, he started to grow again and he was having a kind of rebirth just now. The music he played on his show these days he played because it turned him on. To some extent he was a musical prostitute, but he looked for soul and good style in what he played and he felt he could make changes smoothly and assert his personality without being overly self-indulgent as he had been when he started. But even now he worried about losing touch. He could not be sure it would be good next week just because it was good this week. But at least he was out of the jungle. He could see how someone like Cole was not out of it yet. Cole was not yet sure what he was doing with his personality on the air. Ponek said Cole reminded him of himself a few years earlier, that was how he could tell.

Over the time Ponek had been with the station, a lot had changed. His feelings about promotion men had changed. At first when promotion men came around and thanked him for playing their albums and said it was helping sales, he had felt he was a real economic influence. But after the initial ego inflation, he realized he was not important, his job was important, and the promotion men were really not his friends. With promotion men, you got into a no man's land. You had to be kind to them. They did favors for you, you did favors for them, and both of you felt insecure. You went through a period of playing the favors game. But if you were powerful, if you felt more secure because of a position of real strength, you knew they would not say no if you asked them for something. Some of the promotion men who came to the station were insulting, they made your relationship with the artists secondary to your relationship with them. But some of them did it well, they came on straight and did not insult you and did not take it personally when you said no. Ponek thought maybe three out of five promotion men who came to the station now did not know how to deal with the staff. It was the staff's job to tell them how, and they did, but they did not use to have that attitude.

Simmons felt the station's relationships with promotion men were basically relationships with the record companies and they were fragile and important to maintain. Sometimes the difference it made was two hours, whether you got a Rod Stewart album before a competitor. But the promotion men who were their friends really were. It was a circle, the music business. Simmons felt she was in it, her old man was in it and all her friends. This station, this place, was like a giant succulus on your life, it was your life, your whole life. It was a small place, there was not much space and they were all very close, they got to know everything about each other.

Simmons, in July, was on the air regularly on weekends and sometimes she did fill-ins. She had been working weekends for about a year. The first time she was on she was scared she would not be able to say anything. Her old man left her alone in the studio and walked out of the building. She was more confident now but she did not think she was the kind of person who had much to say to an audience. She was not like Street or McClay. She thought what she had was a specialized knowledge of records, of the record companies and how they changed, and she knew what was going on at the station. She had learned to do production on Laughlin's talk show and on the Dr. Hip show and she had her own taste in music. She knew it was biased but she thought it was good. She was more into country than some of the others.

Ponek said he felt everybody at the station was an egomaniac. They thought they were a privileged few and smart. They were very verbal people, they could talk, and they made fun of other people, outsiders, for not knowing their ways. They had to, they had an image to maintain. The outsiders — the public, the audience, the advertisers, the promotion men expected it. He felt that the station added strength to the individuals who stayed with it, they had to have strength to survive.

McClay said what he cared about mainly was that the station be a happy place. He would not get into it politically until he saw it becoming an unhappy place or until he was interfered with. He thought maybe they had more security with Duff than with Donahue as general manager. Duff had been a company man but he had grown to understand the radio station. Donahue was more transitory. As soon as he came you knew he was going to leave, he got bored.

Donahue said he felt these things had their ebb and flow. With Duff they had ebbed. Now he had to pull it back together. He felt if he let them, the staff would slip into their s.o.p.'s and everyone would get lazy and just do their own part with least effort. They would treat it like a job. It would all boil down to their own separate egos. Donahue thought about the staff a lot. There was Cole and Gossett. Donahue felt he understood them and why they played what they did and that Gossett had the best ear for putting together rock music of anyone they had at the station. Both Cole and Gossett had grown up in San Francisco. Donahue felt they were young, they were in their early twenties, and pragmatic. He said he kept telling Cole, "Bobby, you have to work hard, for the teenager in me." There was Simmons. She was invaluable to him. She would work at the station all the time if her old man did not take her away sometimes. She liked a certain kind of music but like with Street, Donahue felt she should have command of more. There was Street, he was not sure what he could do for Street.

There was Laughlin. Laughlin had been a salesman at KMPX. He was strange, eclectic, psychedelic right wing. Donahue felt the function of Laughlin's talk show on Saturdays was really therapy. It was so all the weirdos in the audience could call in and say what they thought on the air and have him say, "yeah man, right on!" or something like that. There was Melvin. Melvin was still in Nepal but he might be coming back. He had been Donahue's first salesman at KMPX and he was the most like a first lieutenant Donahue felt he had ever had. But Melvin would not stay even if he came back, for all the reasons that made Melvin what he was. There was O'Hair. Duff had hired him. Donahue felt he would like O'Hair to take some things away from him, to come and say he would do them, but O'Hair would not. He was not really a good program director. He could go for several days and forget to say his name on his program.

Then Laufer, he lent a sense of humor to the news. He counterbalanced McQueen. Donahue said once someone called in to complain that a dentist had died by breathing too much of his laughing gas and Laufer covered the story and at the end of it asked if there was a smile on the man's face. But Laufer was irresponsible, he left McQueen too much work. Donahue respected McQueen for his work and he valued McQueen's judgment, but he felt McQueen was lazy. He thought McQueen could be a leader. But McQueen was such a down person. A leader had to have faith, he had to be able to believe that something would work and get other people to believe it too.

There was McClay. Donahue said he always said McClay was pressed off-center. He had known McClay a long time. He had learned to filter out a lot of what McClay said but he listened to him because every once in a while McClay would say something good. He would jump on McClay for his shows, to keep him from slipping. If he kept after him they would be all right. McClay knew how to mix music. There was Bear. From the beginning at KMPX, Donahue felt Bear had potential, he still felt he did. But Bear reeled forth this great philosophy. He had a good relationship with some of the audience and he talked about how they should all touch each other lovingly, but he was a hypocrite. There was Ponek. Donahue did not understand Ponek. He felt Ponek had learned from the group from KMPX but that Ponek really did not have the instinct for it. Ponek would sometimes mix things too obviously, like playing sex sets. Every once in a while he would play a set that was one sex number after another. Donahue said then he would have to tell him, "Now Stefan, we can play each one of those records but they are controversial because of the context, so why string them all together and make it stand out like a sore thumb?"

There were changes Donahue felt he wanted to make at the station: changing the titles of people, having them switch shifts, having the air staff rate cuts from one to 10 on a rating sheet on each album as a possible guide for each other, arranging for everyone, including the secretaries, to take a trip, maybe to Hawaii, teaching the salesmen to write good English. The salesmen were a problem. Donahue felt he always had tension with them. It produced resentment, but part of it was necessary. He had to educate them to new ways of doing things. If there was no resentment he was just not doing it. The salesmen had one kind of relationship with Duff. They had expected Duff to cut things and adjust them, but he was not going to do that. Donahue said he told the salesmen in the beginning not to go after offensive sponsors and not to give something away when they did not have to, and they did not have to. Then there was the problem of death, the death of the radio station and of individuals. Donahue felt he needed Gossett and Cole, the younger ones. When Gossett started, the phone calls he got were mostly from people his age. Now the calls were from younger people, at least they seemed younger, actually he was getting older.

Donahue thought the ideal would be not to have to tell people what to do, but to have them ask to do what they thought they wanted. He wanted them to come to him with ideas, he wanted to be able to say yes, he could do that. As program director, three or four years ago, he had not been able to. Now as general manager, he thought he might, but he was realizing he had to say, "I'll try." He had lots of theories about this. It was not business really, it was psychology. He looked for patterns all the time, in people, in situations. He always had ideas, so many ideas.

Toward the end of July, a new chief engineer with the station, How Wachspress, proposed to Donahue that they have an engineering meeting with the staff. He wanted to discuss the bad state of the equipment and what would get priority. On the afternoon of Thursday, July 27, not the entire staff, but most of them who handled the equipment, met in the Lizard Lounge. It was their first large meeting since June 20 when Duncan came out to fire Donahue. This time Donahue was present. Wachspress began:

Wachspress: We're going to talk about six things and then you're going to say whatever you want. First thing I'm going to talk about is logs.

Street: What?

Wachspress: Logs, and the legal bullshit of running the station. You know, we got to deal with the F.C.C. For the most part everything is kind of, you know, it slides, but still there's the law, and things got to be done, they have to be done, and two things about logs that have to be done. One, they have to be taken, and the second thing is that periodically everybody should really look at the instruments, not to know what the readings are, but to know if the station is on the air.

Street: The freq meter is always off, I mean, it's like two or three points off, it always is, it's never calibrated to zero as long as I've been here.

Wachspress: I calibrate it every day.

Street: You put it on calibrate every day?

Wachspress: Every two days or something.

Hamilton: Wait a minute, you can only calibrate so close... .

Wachspress: It doesn't matter, we happen to have a servo-mechanism that controls the crystal so our frequency is very stable, that's the least of our problems... . What I'm talking about is serious things, like the arcing problems that occur when the fog comes in, when the transmitter is running too hot, or if something is wrong, we have too much current, or something else like that. Everybody who works and has a third-class phone is licensed theoretically to be able to make certain minimal type of adjustments until they can call the first class operator. But there's a whole lot of people that don't know it and that's one of the things that they're going to have to do.

Midway into the meeting, Donahue took the opportunity to let the staff know how he felt not so much about the equipment as about the station and their position with respect to the corporation. His remarks were interrupted periodically by comments from the staff:

Donahue: How, can I say something?

Wachspress: Sure.

Donahue: Maybe not everybody knows about the background. We started off with a guy who was a good engineer named Cassidy. He hated all of us. But he was a pretty good engineer. Then we came into our next engineer who was a half wit. Then during the period of time when Kibler was here, Kibler was one of those guys who was not perfect in a lot of areas but he also did a lot of things to clean up, to get rid of a lot of the crap that had gotten in the way. Kibler was a long way from perfect man, I was one of the first to jump on him for a lot of things, but he did clean up a lot of crap. Now we are trying to get How in to clean up the rest of it.

What we're aiming for is (a) to get our signal going in the right direction, we're very heavy in the ocean and the desert.

[Laughter from staff]

And what we are going to do is to get our signal in a direction that will really do us some good, and then we are going to work to try to bring the signal and the sound of the station up to what I think programming quality can be.

In other words, excellence is what we're shooting for in every area. And what you've got to do when we talk about it is sort of forget about how terrible it's been and let's see how good we can make it, because we can talk all day about the rotten fucking equipment we get around here, and the terrible shit we have to work with, but we are now in the position where I've gotten the okay on going ahead with the new production studio equipment.

Boucher: Oh yeah?

Donahue: Yeah.

Boucher: Out of sight. How much money?

Donahue: Well, you know, we've got enough to do this.

Simmons: Seven dollars.

Boucher: How much?

Donahue: A couple of hundred dollars and a little trade. We've got enough that we're going to be able to come up with a good production studio. It will not be in this year, let us say, the quality of a record plant or something like that but it's going to be much better equipment, much improved equipment, and I will consistently work... let me tell you something, we are getting so that the rest of the radio station is working all right. Okay. We had a profit last week of $6,147. Now that's one of the best weeks, if not the best week, the station's ever had. Just to give you an idea of where we're at at the moment, the total profit —

Ponek: When you're talking about profit, is this business on the air or is this business on the table?

Donahue: I'm talking about profit after commissions are paid, after amortization and depreciation.

[Conversation among the staff]

Donahue: The general point is that, let's say that the radio station was budgeted this year to theoretically make a profit of about $120,000. Up to and through the month of May our profit was $10,000, which only gives you a mere $110,000 to make in the other seven months. But our profit in June was $8,000. And I think we are well on our way now to making the budget. If we make the budget man, we can do all the things we want to do. Already the fact that we've shown as much progress as we have is why we're able to get the production studio equipment, you know, and we're really heading in the right direction, things are working well.

They're working well enough that I am in the position when they call me up with their ridiculous ideas to tell them to go fuck themselves. I told one p.o. in the company today, I said hey, we're supposed to be making a profit and getting ratings, you may or may not be worried about its being good radio also but it is that, and once we have done those things we are doing what we want. Don't come around now and ask me to join your fucking church because I'm not going to join, and I'm not going to change the way all of you would like to have it to get that. They understand that, you know, and we will be increasingly left alone, I'm refusing to go along with a lot of their asinine ideas because they're — but there are also some good ideas.

So this to is my way of thinking part of our attempt to progress, as far as our attempt to progress is making the station sound better. Now I don't think there's anybody here that can't learn more about what goes on in there [the studio], and it is an area, engineering is an area where a little knowledge is really a dangerous thing, because nobody, people can become an expert so fast once they learn to turn three knobs and push a switch, and I think we should try to address ourselves to Mr. Wachspress' sense of what we can learn from him, what he can do for us, and that we can really learn things from him, and that we can really improve the sound of the station.

Wachspress: Like I said to you, I said no matter how much the audience likes us, if they can't get us they're going to turn us off, and what we're selling is sound, we're selling audio.

There followed a discussion by Wachspress and the staff about problems associated with the equipment:

Wachspress: We've got all sorts of crazy loads and distributions... .

We've got the worst type of problems you can imagine, we have intermittents, problems that come and go, they may kick out for a few minutes, a few hours, and then they'll restore themselves, and then they'll come back... .

In Studio One [the small front studio], that Ampex machine has a lousy head, besides which it's probably lousy in a lot of ways which I don't even know of. Before Kibler left he fixed it. Kibler has fixed the machine God knows how many times and it craps out no matter what. That machine is a disaster. It's going to either have to be replaced or have an extensive overhaul. I don't know how the other one is... .

One of the interesting little phenomenological things about this place is that all our monitor speakers short of the ones in the master are mounted in no baffles. It's just simply like grilled cloth and it's just kind of stuck in there. In production, they're mismatched speakers, they're grade x or something.

Boucher: I have mismatched ears.

Wachspress: Let's just do it studio by studio. The board there, well let's not talk about the board... .

Street: If it was just cleaned it wouldn't be so bad but there's so much crap in it... .

Wachspress: Speakers have to be mounted in proper baffles... . I think that, to put it bluntly, we should be able to have some idea of what the hell we're putting out over the air. Right now we're having a very strange idea. I think everybody else knows better what we sound like than we do.

Street: Yeah, right... .

Wachspress: Also, there's an order for some new cue amplifiers and monitors.

Street: Is Macintosh way out of our range?

Wachspress: Actually, quite frankly, I think we should buy SAE equipment.

Boucher: What about Op Amp?

Street: We can get a cut rate on it on the Peninsula.

Wachspress: I don't know... . Let me just lay it right on the top, we're working with a board that's fifteen years old, was originally a mono board, was turned into stereo. It has no audition on it, it has a half-assed cue system. We have no means of auditioning records while we're on the air.

Boucher: Where's this?

Wachspress: Where you work. When Bill Kibler was working here he was working under the old regime where it was like bandaids and solder, do it as cheaply as you can.

Street: Yeah, right.

Wachspress:	My feeling is its's going to be absurd —
Street:	To pour money into it that way —
Wachspress;	Because we're dealing in an age, we're selling audio, when people who are going to go out and spend $500 or $1,000 for a tuner, or even $200 for a tuner, are expecting to hear a certain degree of sound, and the thing is that we can't, we're not delivering it. There are a lot of things that are going to have to be changed, a lot of things. I just got my budget today.

When the meeting was over, Donahue walked out toward his office. He seemed anxious and hopeful and was wearing blue jeans and a blue denim shirt on which someone had embroidered a few small, brightly colored flowers. McQueen stayed in the Lizard Lounge talking with the public affairs director about how the news department had inferior production equipment. He said they had better equipment at KNEW and KNEW was losing money. There was better equipment at KSAN for commercials. Wachspress hung around for a while listening to them and then said he had other priorities. Hamilton went out and started talking with Street about the women's special they would be doing on Sunday. They were calling it, "Women and Music, 'He Hit Me and It Felt Like A Kiss'," after the Crystals' hit.

Conclusion

The dilemma of cooptation is a moral one. When the people of the radio station worry about being coopted, they worry about becoming corrupt, going bad, selling their souls as they sell their time. To some extent their worry is a product of organiational circumstance. In the preceding account, the concern with selling out arises early and persists. It is one kind of response to loss and change and, in particular, to a situation in which losses are also choices. The people of the radio station do in fact give their past away. They make decisions which over time enable the station to take on a broader range of commitments. There is loss as this happens. The organization they started with no longer exists and some of what they prized about it, the initial naiveté, for instance, is sacrificed to memory. There is also gain as people involved and organization become more able.

If the preceding account is to be believed, there is no alternative if the station is to survive than for it to change in such ways as will allow an increasing integration with a larger society. There are, however, different ways of making accommodation. In the histories of other stations, we see some of the routes not taken by this one and in the histories of different individuals with this station, we see a range of ways of coming to terms with the loss and change dilemma. This is extremely important. For the process, while it has imperatives, also has room for discretion.

In tracing the history of the station, we have seen some of how its larger process of change takes shape, some of the room for choice it allows, and some of what the concern with selling out refers to. We have also seen how standards of conduct are relevant to particular times and how judgments of what is necessary vary, as do judgments about what is truly corrupting. Donahue at one point comments:

> There will always be a segment of your audience that considers any kind of advertising that goes beyond head shops and boutiques as a form of sellout. If this is the guide line then we have indeed sold out, as has everyone but The Fool on the Hill (when he stays on the hill). I do not believe the commercial per se is evil.

The distinction is possibly crucial.

This story of the radio station is one account of a cooptation process, one which in its detail is more instructive than in any generalization that can be drawn from it. The following, however, are some summary statements which may be made. Part I (Beginnings): In the very beginning, people come together to set up the new station, working out agreements in terms of relatively simple exchanges, making tentative, short-term commitments. They act in accord with institutional and personal contexts, assuming roles that are relevant and known: disc jockey, owner, investor, salesman, advertiser, listener. They gain space for the station gradually and in a series of trials define its style. Different kinds of interest in the station soon feed each other. As the station and its audience become known, advertisers get results, rates can be raised, income increased, more listeners gained. There is confirmation for early investment and encouragement to do more. The very facts of success are dramatic enough for public story.

Part II (Legitmacy): The staff now must learn to operate in the context of a large corporation. There are new rules with which they are expected to comply and these are made clear in introductory meetings, new appointments, and corporate attempts at control. Disillusionment of the staff is especially noticeable during this period. It accompanies efforts to broaden the station's pattern of support by appealing to listeners and advertisers unfamiliar with the station's style. This in turn exacts concessions from that style. Other stations with similiar formats soon enter the market and the competitive situation is changed. Yet the fact of their emergence indicates an expanding market for the format and contributes to the station's legitimacy.

Part III (Professionalism): This is a time of search for standards of conduct that will enable the station to function successfully in the realm in which it is increasingly implicated. There is in particular a concern for rules which will separate personal preferences, political views, and outside interests from what is appropriate to the station. The sense in the end is of having resolved the underground/establishment dilemma in a way more clearly favourable to the establishment than had been apparent previously. What comes to be prized especially by this time is a competence in dealing with the world which, while serving the station, is not specific to it. The nature of this competence is indicated in discussions of professionalism.

Part IV (Renewal): A problem suggested by criticism of the station during this period is that of isolation. Internal distinctions made by the staff to reconcile their different commitments are often not comprehended in assessments made by others and one result is that the station seems, and is, to a large extent inaccessible, but for different reasons now than earlier in its history. There is a concern that the staff not lose touch with the audience and parts of themselves that may be important to the station's survival. There is also repeated admonition that they realize what they are doing and not do it blindly.

The process continues beyond the period of study. The radio station of the present account is successful in its market six years later. It continues to be criticized for having sold out. Yet it has not entirely abandoned its original style. Disc jockeys still select their own music. The news has a curious autonomy. There are, however, some significant changes. Donahue dies of a heart attack in 1975 and the loss is profound. The manager who replaces him is more ordinary. Simmons becomes the program director and midday jock and is central to the competence of the station. Yet by the summer of 1978, most of the old-timers, including Simmons, have left. The community is not there for them like it used to be. There is a new generation listening who are not loyal, who have not had the experience of the 1960s and the Vietnam war. The audience as a whole is a little older, heavier in the 25-34 range. The market is splintered by the presence of 10 other FM stations which play different special varieties of rock. The station, to compete, plays a higher proportion of hits. There is more advertising on the air in general categories, more airlines and banks, a moving company. The studios are located in the heart of the financial district where the station now has a separate building.

Yet even in 1972, it was a long way from the initial enthusiasm and the radio station that could never happen unless someone inherited one out of the blue. The process of cooptation with which this radio station was involved is in part a tragic one. There are many small deaths along the way as one habit of doing things is replaced by another and these over time have large consequence. Donahue in the final chapter replies to a question from a listener who asks about underground radio and why it is dead: "Because life is a series of moments man, that you live while they're happening and don't expect to constantly repeat them." Yet it is worth remembering that one measure of the present is the past and sometimes, as Bear suggests, the present seems "lame and untrue to the spirit of it all." There really is such a thing as grief and although the history of a radio station may be about commercials and waterbeds and hiring and firing people who play records, it is also about how all of us do, as time goes on, leave behind things that matter, and about how despite the failures and regrets and the changes of a larger world, the steps continue to be ours.

Sources

In the notes which follow, interview and documentary sources are given in order of appearance for each chapter. Interview sources are listed first followed by unpublished documents and publications. For the unpublished documents, the person or organization in whose possession they were found is indicated in parenthesis. I am grateful to all those named as interview sources for their willingness to speak their minds and for their very admirable belief in the value of what they were doing. The interviews were conducted during May 1972-April 1973. Their time-boundedness should be noted, as people often change their minds and later see things in a different light.

Publications frequently referred to have been abbreviated as follows: (BG) San Francisco *Bay Guardian*; (BB) Berkely *Barb*; (BT) Berkeley *Tribe*; (Bbd) *Billboard*; (Bcstg) *Broadcasting*; (DOB) San Francisco *Dock of the Bay*; (ET) San Francisco *Express-Times*; (GT) San Francisco *Good Times*; (RS) *Rolling Stone*; (SFC) San Francisco *Chronicle*; (SFE) San Francisco *Examiner*; (SFEC) San Francisco *Examiner and Chronicle*.

Other abbreviations are: (MR) monthly reports in the possession of KSAN, preceded by name of month. The year of each monthly report is indicated in the title of the chapter in which it appears unless otherwise noted. (Mm) Metromedia, Inc., corporate owner of KSAN. (Bio) biographical information obtained from the station or individuals.

Introduction

1. Nicholas von Hoffman, "The Acid Affair — XIII," *The Washington Post*, Oct. 27, 1967.
2. Michael Rossman, "KMPX On Strike," *Express-Times*, Mar. 21, 1968.
3. The classic study of cooptation in sociological literature, to which much later work refers, is Philip Selznick, *TVA and the Grass Roots: A Study in the Sociology of Formal Organization*, New York: Harper Torchbooks, 1966, Berkeley and Los Angeles, University of California Press, 1949. According to Selznick, "To risk a definition: cooptation is the process of absorbing new elements into the leadership or policy determining structure of an organization as a means of averting threats to its stability or existence." While Selznick looked at the process from the point of view of an organization doing coopting, the present account examines an organization coopted.
4. Eighty-eight people were interviewed. Fifty-four of them had worked for the station at one time or another over the five years. The documentary evidence obtained included station monthly reports, memos, letters, program schedules, promotion materials, legal and

financial records, magazine and newspaper articles, rating service audience estimates, and publications dealing with radio and the rock music industry. There were also, although fewer in number, tapes of programs, commercials, and meetings.

5. Susan Krieger, *COOPTATION: A History of a Radio Station*, Ann Arbor, MI:Xerox University Microfilms, 1976.

6. This is discussed in Susan Krieger, "Research and the Construction of a Text," *Studies in Symbolic Interaction*, Vol. 2, 1979.

7. Elinor Langer, "The Women of the Telephone Company," *New York Review of Books*, March 26, 1970; also in Dorothy Richardson, (ed.), *Women at Work*, Chicago: Quadrangle, 1972.

PART I: BEGINNINGS

Interviews: Donahue, Crosby.
Documents: notification of format change, KMPX, Dec. 5, 1968 (FCC); license renewal application, KMPX, Aug. 26, 1968 (FCC); station records, KMPX (Crosby); financial accounts, KMPX (Shade for Crosby).

Chapter 1: KMPX (March-June 1967)

Interviews: Donahue, Boucher, Crosby, Hunt, Yochim, Talbot, Hamilton, Miller, Ickes, Gleason, Harris, Fong-Torres, McClay, Johnson, Prescott, Melvin, Nelson, Wehr.
Documents: depositions of Donahue and Crosby in Janet Cook v. Crosby Broadcasting Company et al., Superior Court, San Francisco, No. 601, 589, May 19, June 2, 1970 (Phillips for Crosby); financial records (Crosby); letter, Hyde to Donahue and Mitchell, Dec. 8, 1966 (KSAN); North Face fact sheet (Hamilton); bio, Prescott (KSAN); letter, SL and BC to Underground, Apr. 26, 1967 (Hamilton).
Publications: panel discussion, "The Rock'n'Roll Industry," *California Monthly*, University of California, Berkeley, June-July 1967.

Chapter 2: KMPX (April-November 1967)

Interviews: Rogers, Donahue, Harris, Melvin, Boucher, Wehr, Johnson, Street, Hamilton, Voco, Towle, Laughlin, Talbot, Tagnoli, Gleason, Crosby.
Documents: letter, Feldman to Crosby, Apr. 14, 1967 (Crosby); KMPX Rate Card No. 6, May 15, 1967 (Harris); KMPX advertising records, April-July 1967 (Hamilton); letter, Bead Freak to KMPX, July 1967 (Hamilton); telegram, Gramophone to KMPX, June 28, 1967 (Hamilton); letter, Wehr to KMPX, Aug. 7, 1967 (Hamilton); letter, Wangeman to Donahue, June 23, 1967 (Hamilton); reply, Donahue to Wangeman, July 3, 1967 (Hamilton); listener letters and Donahue replies, April-July 1967 (Hamilton); Donahue-record company correspondence, June-July 1967 (Hamilton); North Beach payroll records (Hamilton); advertising agency lists, Melvin, Oct. 1, 1967 (Hamilton); letter, Young to KMPX, Sept. 15, 1967 (Hamilton); August-Sept. Pulse summary, KMPX (Hamilton); monthly business accounting, July 1967-Jan. 1968 (Crosby); sales gross summary, Sept. 1967-Mar. 1968, Donahue (Hamilton); financial notes, Crosby (Crosby); financial notes, Avery (Rogers); KMPX complaint, file per WBR, Complaints and Compliance Division, FCC (FCC); letter, Silverman to KMPX, July 9, 1967 (Hamilton); letter, Rodes to KMPX, July 27, 1967 (Hamilton); letter, Holder to KMPX, Sept. 29, 1967 (Hamilton); reply, Hamilton to Holder, Oct. 26, 1967 (Hamilton); memo, Donahue to All Announcers,

Oct. 9, 1967 (Hamilton).
Publications: Ralph J. Gleason, *The Jefferson Airplane and the San Francisco Sound*, N.Y.: Ballantine, 1969; Nicholas von Hoffman, "The Acid Affair-XII," *The Washington Post*, Oct. 27, 1967; "Until just recently . . .," *Sunday Ramparts*, Apr. 23, 1967; "KMPX: 106.9 mg stereo . . . ," *Mojo-Navigator*, Apr. 1967; "KMPX Brings The New Sound Into Your Home," George Washington High School *Eagle*, June 1967; Gleason, "Tom Donahue is on KMPX-FM . . . ," *SFC*, Apr. 12, 1967; Richard Barrett, "No Hex on KMPX," *BB*, July 28, 1967; Gleason, "Something New on the Pop Music Scene," *SFC*, Aug. 16, 1967; Gleason, "Perspectives: One Hundred Rolling Stones," *RS*, Jan. 20, 1972; Bob McClay, "Murray the K on WOR-FM, 'They Screwed It Up,' " *RS*, Nov. 9, 1967; Tom Donahue, " 'A Rotting Corpse Stinking Up the Airways . . . ,' " *RS*, Nov. 23, 1967; Eliot Tiegel, "Free Form Show Is Kicked Off by KPPC," *Bbd*, Nov. 11, 1967; "SF 'Hippop' Music Format of Future?" *Bbd*, June 24, 1967; Claude Hall, "Progressive Rock All the Way In WNEW-FM's Format Future," *Bbd*, Dec. 9, 1967; Claude Hall, "KMPX-FM's Donahue Programs Music With a Wide Open View," *Bbd*, Dec. 30, 1967.

Chapter 3: KMPX and KPPC (November 1967-March 1968)

Interviews: Bear, Johnson, Melvin, Harris, Weinberger, Hunt, Shade, Hamilton, Donahue, Prescott, Crosby, Boucher, Street, Miller, Stone, Gleason, Voco, Tagnoli, Rogers, Ickes, Laughlin.
Documents: bio, Bear (KSAN); bio, Harris (KSAN); KMPX Rate Card No. 8, Jan. 1, 1968 (Harris); memo, Hunt to Staff, Feb. 12, 1968 (Prescott); note, Miller to Whoever (Prescott); memo, Donahue to Miller, Jan. 29, 1968 (Hamilton); memo, Donahue to Prescott (Prescott); notes of meeting, CAB, Jan. 16, 1968 (Rogers); financial statements, balance sheets, CAB, Dec. 31, 1967, Jan. 31, Feb. 29, 1968 (Rogers); letter, Rogers to Avery, Feb. 12, 1968 (Rogers); notes of meeting, CAB, Feb. 15, 1968 (Rogers); financial records (Crosby).
Publications: "Larry Miller and KMPX-The falling out," *ET*, Feb. 15, 1968; Sandy Darlington, "Cream," *ET*, Mar. 14, 1968.

Chapter 4: The KMPX Strike (March-May 1968)

Interviews: Donahue, Prescott, Johnson, Hughes, Pigg, Harris, Crosby, Bear, Hunt, Boucher, Melvin, McClay, Towle, Laughlin, Miller, Street, Kessler, Bird, Ickes, Graham, Weinberger, Wehr, Yurdin, Ponek, Dunlop, Leath, Sullivan, Duncan.
Documents: tape, prestrike program, KMPX, Mar. 18, 1968 (Boucher); "Notice!!," AAFIFMWW, North Beach Local No. 1, Mar. 18 1968 (Harris); "The Tribe of People . . . ," AAFIFMWW, Mar. 20, 1968 (Harris); list of strike events, Harris (Harris); news release, AAFIFMWW, Mar. 20, 22, 1968 (Harris); cable, Stones to AAFIFMWW, Mar. 21, 1968 (Harris); notice and notes of meeting, CAB, Mar. 21, 1968 (Rogers); affidavit of Larry Ickes, Apr. 29, 1968, in charge against AAFIFMWW filed by Leon A. Crosby, May 3, 1968, NLRB Case No. 20-CB-1843 (Rogers); note of telegram, Crosby to Avery, May 4, 1968 (Rogers); note of telegram, Crosby to Avery, May 6, 1968 (Rogers); KMPX advertiser list, AAFIFMWW (Harris); letter, Leath to Dougherty, June 12, 1968 (KSAN); annual reports, Mm Inc., 1959-1968 (Mm).
Publications: Jef Jassen, "KMPX Walk Out Beautiful," *BB*, Mar. 22, 1968; George Gilbert, "Rock Radio in the Streets," *SFC*, Mar. 19, 1968; "A Strike at Rock Station," *SFC*, Mar. 18, 1968; "The staffs . . . ," *AP*, Mar. 18, 1968; Gilbert, "Rock Radio in the Streets," *SFC*, Mar. 19, 1968; Darlington, "KMPX on Mars," *ET*, Mar. 28, 1968; The Old

Ranger, "Free Rock Radio Strike," SF State *Daily Gator*, Mar. 27, 1968; Gleason, "Scorpions, Destiny and a 'Hippie' Strike, *SFC*, Mar. 20, 1968; William O'Brien, "A Fruminous Benefit for KMPX Freedom Fighters," *SFE*, Mar. 21, 1968; Michael Rossman, "KMPX On Strike," *ET*, Mar. 21, 1968; "Digging FM Rock," *Newsweek*, Mar. 4, 1968; Darlington, "And The Winner Gets The Radio Station," *ET*, Apr. 11, 1968; Superball advertisements, *BB*, Mar. 29, 1968, *ET*, Apr. 4, 1968; Jassen, "Larry Miller Throws Support to KMPX Strikers," *BB*, Apr. 5, 1968; Jassen, "Of Stryx and Shux," *BB*, Apr. 12, 1968; Jassen, "KMPX Strikers Credit Barb With a Win," *BB*, Apr. 19, 1968; Jassen, "Larry Miller Scabs Again, Yet, of Course," *BB*, May 3, 1968; Jerrold Greenberg and Ben Fong-Torres, "KMPX-KPPC Shut Down: FM Workers Strike for Rights," *RS*, Apr. 27, 1968; Gleason, "They All Came Out for KMPX Strikers," *SFC*, Apr. 5, 1968; Catessa, "Letters to the World, KMPX," *ET*, Apr. 18, 1968; KMPX advertiser list, *BB*, Apr. 19, 1968; "The Man Who Buys Red Ink," *Forbes*, Feb. 15, 1967; "A'bum' year for a 'glamorous' industry," *Forbes*, Jan. 1, 1968; "Ad Complex Raises Sights," *Financial World*, Mar. 8, 1967; "Metromedia-Profitable package," *Financial World*, Apr. 17, 1968; Jassen, "Now It's KSAN Reader-SAN," *BB*, May 17, 1968; Lenny Lipton, "At the Flick," *BB*, May 17, 1968; Anathin, "KMPX Strikers Get New Station," *ET*, May 16, 1968.

PART II: LEGITIMACY
Chapter 5: KSAN (May-August 1968)

Interviews: Sullivan, Hamilton, Donahue, Prescott, Campbell, Ponek, Powell, Kibler, Kaffen, Leath, Dougherty, Paulsen, Pigg, Boucher, Stone, Duncan, Harris, Klein, Melvin, Laughlin, Dunlop, Towle.
Documents: music list, Donahue, May 21, 1968 (Hamilton); letter, Ray to Dougherty, May 23, 1968 (Dougherty); letter, Ray to Dougherty, June 5, 1968 (Dougherty); reply, Dougherty to Ray, June 19, 1968 (Dougherty); telegrams to Donahue, May 21, 1968 (Hamilton); July MR; memo, Paulsen to Staff, July 22, 1968 (KSAN).
Publications: Gleason, "Hippie Strikers Back at Work," *SFC*, May 22, 1968; "KSAN-FM to Progressive Rock," *Bbd*, May 25, 1968; Robert Commanday, "The Pollution of the Airwaves," *SFC*, June 4, 1968; KSAN ad, *BB*, May 31, 1968; Darlington, *ET*, May 23, 1968.

Chapter 6: KSAN (September 1968-March 1969)

Interviews: McClay, Melvin, Voco, Hughes, Street, Harris, Sullivan, Laughlin, Donahue, Duncan, Klein, Powell, Bear, Hunt, Ickes, Crosby.
Documents: Melvin wedding invitation (Hamilton); Oct.-Jan. MR's; memo, Paulsen to Entire Staff, Feb. 11, 1969 (Harris); memo, Bear to Friends, Feb. 21, 1969 (Harris); memo, Bear to Harris, Feb. 10, 1969 (Harris); Donahue, "Seeds and Stems," draft, Jan. 3, 1969 (Hamilton); letter, Phillips to O'Reilly and purchase agreement exhibits, Feb. 25, 1969 (FCC); memo, Paulsen to Staff, Mar. 27, 1969 (KSAN).
Publications: "Family Radio," *ET*, Aug. 28, 1968; "Hip Travus Hipp," *BB*, Oct. 18, 1968; Claude Hall, "Progressive Rock Stirs Radio World Across U.S.," *Bbd*, Oct. 19, 1968; "Another Merger Faces FCC," *Bcstg*, Oct. 14, 1968; "KSAN Sacks Bear," *BB*, Feb. 14, 1968; "Edward Bear-No Soap, No Radio," *ET*, Mar. 11, 1969; Donahue, "Seeds and Stems," *Record World*, Feb. 17, 1969.

Chapter 7: A Voice of the Revolution (April-August 1969)

Interviews: Donahue, Hamilton, Powell, Duff, Prescott, Harris, Paulsen, Campbell, Pigg, Ponek, Klein, Dunlop, Gardner, Nisker, Voco, McClay, Melvin, Bear.
Documents: bio, Duff (KSAN); Apr. MR; bio, Ponek (KSAN, Ponek); bio, Campbell (KSAN); letter, Gardner to Art Blum Agency, June 24, 1969 (Gardner); letter, Gardner to Kornfield, June 19, 1969 (Gardner); letters, contracts, commercial copy, Gardner, Woodstock Excursion, June-July 1969 (Gardner); July, Aug. MR's; May MR; memo, Duff to Ponek, May 1969 (Ponek); memo, Ponek to Staff, May 1969 (Ponek); letter, Melvin to Local Board, July 24, 1969 (Melvin).
Publications: Gleason, *The Jefferson Airplane*; Donahue wedding photo, *RS*, May 31, 1969; Roger Rapoport, "The Underground Radio Turn-On," *Look*, June 24, 1969; Gleason, "The Woodstock Festival . . . ," *SFC*, Aug. 30, 1969; "The New Respectability of Rock," *Bcstg*, Aug. 11, 1969; Donahue, "Metanomena," *Cashbox*, Aug. 23, 1969; bob o'lear, "Medium Hot," *GT*, Aug. 21, 1969; o'lear, "KSAN, luke warm about medium hot," *GT*, Aug. 28, 1969.

PART III: PROFESSIONALISM
Chapter 8: A Voice of the Revolution (September-December 1969)

Interviews: Ponek, Donahue, Pigg, Bear, Prescott, Powell, Campbell, Kaffen Duff, McClay, Harris, Young, Gossett, Dougherty, Boucher, Street.
Documents: Sept. MR; memo, Duff to Steph/Wes, Sept. 15, 1969 (KSAN); memo, Prescott to Cover Cats, Oct. 1969 (Prescott); transcription, Nov. 5, 1969 of Roland Young, Nov. 3, 1969, in Duff to Dougherty for submission to FCC (Dougherty); letter, Dougherty to Ray, Dec. 19, 1969 (Dougherty); statement, Duff, "KSAN has terminated . . . ," Dec. 5, 1969 (KSAN); Arbitration, Decision and Award, in Controversy between IBEW Local 202 and Mm, Inc., Licensee of KSAN-FM, Involving Discharge of Roland Young, John B. Lauriten, Arbitrator, Dec. 18, 1970; Dec. MR.
Publications: "Roland Keeps Rollin'," *BT*, Oct. 31, 1969; o'lear, "We Can Work It Out," *GT*, Oct. 30, 1969; Susan Goodrick, "Roland Young Quits KSAN," *DOB*, Nov. 4, 1969; Ellen Snyder, "KSAN Nowhere to Go," *DOB*, Nov. 11, 1969; interview, Young, by KPFA, in Fred Hampton, "an interview with Roland Young," *Leviathan*, Mar. 1970; Young, "Rollin' Right On," *BT*, Dec. 19, 1969; "Hilliard 'Support' Costs KSAN Disc Jockey a Job," *SFC*, Dec. 6, 1969; Mead, "Media Management," *GT*, Dec. 11, 1969; "Return Roland!" and "Petition," *BT*, Dec. 19, 1969; Leo Lawrence, "Roland Raps, Leo Listens," *BT*, Dec. 26, 1969; Jonathan Eisen, ed., *Altamont*, N.Y.: Fusion, 1970; Gleason, "Tomorrow . . . Sears Point Raceway . . . ," *SFC*, Dec. 5, 1969; "300,000 Say It With Music," *SFEC*, Dec. 7, 1969; "The Love Generation Hype in the News," *RS*, Jan 21, 1970; transcript, Altamont broadcast, KSAN, Dec. 7, 1969, in Eisen, *Altamont*; "Let It Bleed," *RS*, Jan. 21, 1970; Andy Gollen, "People Wonder What Went Sour," *SF Progress,* Dec. 10, 1969; o'lear, "Archangel," *GT*, Dec. 11, 1969; "The Bleak Aftermath of Altamont," *SFC*, Dec. 9, 1969; Donovan Bess, "Angel's Altamont Story," *SFC*, Dec. 9, 1969; Todd Gitlin, "Altamont: end of an era?" *Daily Californian*, Jan. 27, 1971.

Chapter 9: KSAN (January-May 1970)

Interviews: Pigg, Donahue, Turner, Dougherty, Duncan, Croninger, Bensky, Nisker, Duff, Ponek, Prescott, Howell, Boucher, McQueen, McClay.

Documents: letter, Duff to Croninger, Feb. 18, 1970 (KSAN); letter, Mulford to Burch, Feb. 18, 1970 (FCC); memo, Duff to Wes, Larry, Glenn, Feb. 25, 1970 (KSAN); letter, Duff to Carol, July 2, 1970 (KSAN); memo, Duff to Croninger, Mar. 3, 1970 (KSAN); Feb.-May MR's.
Publications: "Radio Tips Off Protestors," *SFE*, Feb. 18, 1970; "Radicals Veterans of Other Riots," *SFC*, Feb. 18, 1970; "KSAN Newscaster Quits," *SFE*, Mar. 2, 1970; "KSAN Explains News Shakeup," *SFC*, May 3, 1970; Scoop, "Spiroed," *GT*, Mar. 5, 1970; "KMPX Sweep," *GT*, Mar. 5, 1970; 'Hip to Hang," *BT*, Mar. 27, 1970; "Four Arrested in KSAN Protest," *SFC*, Mar. 30, 1970; Jassen, "Stop the Wood Shuck," *BB*, May 29, 1970; Bensky, "Ksanitied," *GT*, May 29, 1970; "KSAN KRAK Down," *BT*, June 5, 1970.

Chapter 10: KSAN (June-October 1970)

Interviews: Street, Ponek, McQueen, Boucher, Congress, Harris, Stone, Bear, Pigg, Gossett, Donahue, Fong-Torres, Duff, Bensky, Young, Prescott.
Documents: June MR; bio, Fong-Torres (KSAN); Aug. MR; memo, Harris to Salesmen, July 14, 1970 (Harris); memo, Harris to Boucher, Aug. 31, 1970 (Harris); Sept., Oct. MR's; programming, KMPX, Oct. 1970; tape, McQueen KSAN Newscast, Oct. 21, 1970 (Boucher); press release, KMPX, A Division of National Science Network, Inc., Oct. 20, 1970 (Prescott); memo, Duff to McQueen, Oct. 21, 1970 (KSAN).
Publications: Charles Perry, "The News Reaches SF," *RS*, Oct. 29, 1970; Aloma Sue, "Dusty is an Easy Street," *BT*, July 3, 1970; Fong-Torres, "The Saddest Story in The World," *RS*, Oct. 29, 1970; Karl Marxconi, "Liberated Radio Turnoff," *BB*, Oct. 23, 1970; mike rofune, "off the air," *GT*, Oct. 23, 1970; "Rock Station's Staff Revolt," *SFC*, Oct. 21, 1970.

Chapter 11: A Different Station (November 1970-May 1971)

Interviews: Nemerovski, Sanders, Lavin, Weinberger, Dymond, Duff, Campbell, McQueen, Lambert, Diamond, Bird, Pantell, Wachspress, McClay, Street, Simmons, Donahue, Duncan.
Documents: letter, Duff to McCann, Dec. 9, 1970 (KSAN); Dec.-Feb. MR's; May MR; advertising contracts (KSAN); leaflet, "Vital Information: Care of Birds," Oil Spill Rescue Group, Jan. 1971 (Harris); notes, Pantell, IBEW-KSAN contract negotiations, 1971 (Pantell); IBEW-KSAN Agreements, 1971-1974, 1969-1971 (KSAN); Arbitration, Decision and Award, in Controversy Involving Discharge of Roland Young, Dec. 18, 1970; notes, Pantell, IBEW-KSAN contract negotiations, 1969, 1971 (KSAN); Mar. MR; May MR; letter Knutson to KSAN, Mar. 30, 1971 (KSAN); reply, Duff to Knutson, Apr. 3, 1971 (KSAN); note, Duff to Duncan, (?)May 1971 (KSAN).
Publications: "The Big Oil Spill Spreads," *SFC*, Jan. 20, 1971; Thomas King Forcade, *Caravan of Love and Money*, NY: Signet, 1972; John Grissim, Jr., *We Have Come For Your Daughters*, NY: Morrow, 1972.

Chapter 12: KSAN (June-October 1971)

Interviews: Harris, Dymond, McClay, Duff, Duncan, Boucher, Bear, Street, O'Hair, Ponek, Laughlin, Jacopetti, Johnson, Street, Cole, Simmons, Sanders, Skinner, Gardner, Dunlop, Nemerovski, Campbell, Nisker, Kibler, McQueen.
Documents: June-Aug. MR's; KSAN promotion portfolio, Dymond (Dymond); press release, O'Hair P.D. (KSAN).

Publications: Thomas Albright, "Cartoon Visions on Muni Walls," *RS*, Nov. 11, 1971; "A Boast on Billboard Firm Bomb," *SFC*, Sept. 8, 1971; local press clips, Sept. 1970 (KSAN); "The People Mix Metro's Media," *BT*, Sept. 10, 1971; *SFE*, Oct. 18, 1971; *BB*, Oct. 22, 1971; "To KSAN, Love, Tribe," *BT*, Oct. 22, 1971; "McQueen Comes Clean," *BB*, Oct. 29, 1971; "KSAN KONFAB-McQueen Says He'll Resist," *BB*, Nov. 5, 1971.

PART IV: RENEWAL
Chapter 13: An Incident (November 1971-January 1972)

Interviews: Elliott, Duff, Duncan, Dougherty, Werfhorst, Fong-Torres, Hester, O'Hair, Laughlin, Schoenfeld, Simmons.
Documents: letter, Sugarman to FCC, Nov. 8, 1971 (FCC); reply, Landry to Sugarman, Nov. 15, 1971 (KSAN); letter, Ray to Metromedia (KSAN), Dec. 6, 1971 (FCC); letter, Dougherty to Ginsberg, Dec. 27, 1971 (FCC); letter, Dougherty to Ray, Jan. 17, 1972 (KSAN); memo, O'Hair to Duff, Nov. 19, 1971 (KSAN); memo, Duff to Dougherty, Dec. 16, 1971, draft (KSAN); affidavit, Duff, Jan. 14, 1972 (KSAN); Nov. 1970 MR; Nov. 1971 MR; transcript of program, Dr. Hip Pocrates, Nov. 7, 1971 (KSAN); letter, Duff to Schoenfeld, Nov. 10, 1971 (KSAN); affidavit, Simmons, Jan. 6, 1972 (KSAN); submissions, Dougherty to Ray, Jan. 17, Feb. 23, 1972 (KSAN).
Publications: "The Rolling Stone Papers," *The Realist*, Mar.-Apr. 1972; transcript of program, Dr. Hip Pocrates, Nov. 7, 1971, in Eugene Schoenfeld, M.D., *Dr. Hip's Natural Food and Unnatural Acts*, NY: Delacorte, 1974.

Chapter 14: KSAN (November 1971-April 1972)

Interviews: Duncan, McClay, O'Hair, Harris, Duff, Donahue, Sanders, Hester, Ponek, Cole, Campbell, Gossett, Bear, Crosby, McQueen, Nemerovski, Gleason.
Documents: memo, Duff to Ponek, Feb. 2, 1972 (Ponek); Nov. MR; bio, Nevada (KSAN); Apr. MR; programming, KSAN-KMPX Retrospective, Apr. 8, 1972; staff list, KSAN, Apr. 1972 (KSAN); Dymond, "KSAN's Donahue Presents," press release, KSAN, Apr. 5, 1972 (KSAN).
Publications: John Wasserman, "When The Old KMPX Died," *SFC*, Mar. 6, 1972; "KMPX Turns to Easy Listening," *SFE*, Mar. 4, 1972; moon woman, "KMPX," *GT*, Mar. 10, 1972; "KEMO-TV," *GT*, Mar. 10, 1972; "Random Notes," *RS*, Mar. 30, 1972; Eliot Tiegel, "KMPX Sees MOR as Answer For Ratings," *Bbd*, Mar. 18, 1972; Nina Bernstein, "Is Underground Radio A Bad Trip?" *Variety* (N.Y.), Apr. 5, 1972; "KSAN Cuts 'Gay Report'," *BB*, Feb. 25, 1972; "A Number of Bay Area Women . . . ," *SFC*, Mar. 9, 1972; "KSAN: hip/pig radio," *BT*, Mar. 17, 1972; "Radio Highlights," *SFC*, Apr. 8, 1972; Rich Zimmerman, "Let the Air Waves Flow," San Mateo *San Matean*, Apr. 21, 1972; Gleason, "FM Stations Get Higher," *SFEC*, Apr. 16, 1972; The Staff, "The Truth About Radio," *The Night Times*," Apr. 5, 1972; Wasserman, "Tuning-In To Underground FM," *SFC*, Apr. 3, 1972; Gleason, 'Disc Jockey Credibility," *SFEC*, Apr. 10, 1972.

Chapter 15: KSAN (May-July 1972)

Interviews: Donahue, Bear, O'Hair, Duncan, Fong-Torres, McClay, Street, Campbell, Boucher, Kibler, McQueen, Hester, Gossett, Hamilton, Duff, Cole, Simmons, Voco, Pigg, Dunlop, Ponek, Wachspress.

Conversations among staff: McClay and others, June 21, 1972; Sanders, Gardner, May 12, 1972; Ponek, Donahue, Aug. 21, 1972; Donahue, McQueen, Howell, Wachspress, Hamilton, Street, July 27, 1972.

Documents: tape, staff meeting, KSAN, June 20, 1972; programming, KSAN News, June 21, 1972; May MR; tape, Donahue talk show, KSAN, May 24, 1972; Voco, "Lights Out: San Francisco," Blue Thumb; June MR; sales presentation, "KSAN Delivers," Apr.-May ARB (KSAN); tape, engineering meeting, KSAN, July 27, 1972 (Wachspress).

Publications: Wasserman, "Tom Donahue, The Man . . . ," *SFC*, June 21, 1972; Wasserman, "Serenity Returns to KSAN," *SFC*, June 23, 1972; "Rolling Stones Concert Sold Out," *SFC*, May 15, 1972; Wasserman, "Today is being celebrated . . . ," *SFC*, May 16, 1972; "Underground Television," *The Night Times*, Aug. 9, 1972.

Appendix

CAST OF CHARACTERS

The following is a complete list of persons whose names appear in the text. The position given is the one most often found in connection with each.

Avery, Lew (initial benefactor, KMPX).
Bear, Edward (disc jockey, KMPX, KSAN).
Bensky, Larry (newsman, KSAN).
Bird, Ed (business manager, IBEW Local 202).
Boucher, Paul (engineer and production director, KMPX, KSAN).
Campbell, Donna (business manager, KSAN).
Cole, Bobby (disc jockey, KSAN).
Congress of Wonders (comedy team, KMPX, KSAN).
Croninger, Dave (president, Metromedia Radio Division).
Crosby, Leon (owner, KMPX).
Diamond, Faybeth (oil spill organizer, KSAN).
Donahue, Tom (program director, KMPX; general manager, KSAN).
Dougherty, Tom (chief legal counsel, Metromedia).
Duff, Willis (general manager, KSAN).
Duncan, George (president, Metromedia Radio Division).
Dunlop, Doug (salesman, KSFR, KSAN).
Dymond, Joanne (promotion director, KSAN).
Elliott, Rona (producer, Dr. Hip Pocrates program).
Fong-Torres, Ben (editor, *Rolling Stone*; disc jockey, KSAN).
Gardner, Rick (salesman, KSAN).
Gleason, Ralph (columnist, San Francisco *Chronicle*).
Gossett, Richard (disc jockey, KSAN).
Graham, Bill (producer, The Fillmore).
Hamilton, Raechel (business manager, North Beach Productions).
Harris, Whitney (salesman, KMPX; sales manager, KSAN).
Hester, Mike (producer, KSAN).
Howell, Glenn (public affairs director, KSAN).
Hughes, Lynn (record librarian, KMPX, KSAN).
Hunt, Ron (sales manager, KMPX).
Ickes, Larry (disc jockey, KMPX).
Jacopetti, Roland (production director, KSAN).

Johnson, Katie (engineer, KMPX).
Kaffen, Sue (promotion director, KSAN).
Kessler, Steven (listener at Superball).
Kibler, Bill (chief engineer, KSAN).
Klein, Jerry (sales manager, KSAN).
Kluge, John (president, Metromedia).
Krassner, Paul (talk show guest, KSAN).
Lambert, Paul (oil spill organizer, KSAN).
Laufer, Peter (newsman, KSAN).
Laughlin, Chandler (salesman, KMPX; talk show host, KSAN).
Lavin, Mike (owner, Undulator Waterbeds).
Leath, Reid (general manager, KSAN).
McClay, Bob (disc jockey, KMPX, KSAN).
McQueen, Dave (news director, KSAN).
Melvin, Milan (salesman, KMPX, KSAN).
Miller, Larry (disc jockey, KMPX).
Nelson, Carolyn (media buyer, Jack Wodell Associates).
Nemerovski, Jeff (salesman, KSAN).
Nisker, Scoop (news director, KSAN).
O'Hair, Thom (program director, KSAN).
Pantell, Bob (director of personnel, Metromedia).
Paulsen, Varner (general manager, KNEW, KSAN).
Phillips, Ted (lawyer for Crosby).
Pigg, Tony (disc jockey, KSAN).
Ponek, Stefan (disc jockey KSFR, KSAN; program director, KSAN).
Powell, Lillian (traffic director, KSAN).
Prescott, Bob (disc jockey, KMPX, KSAN).
Rogers, Harry (lawyer for Crosby).
Sanders, Pam (traffic director, KSAN).
Schoenfeld, Eugene (host, Dr. Hip Pocrates program, KSAN).
Shade, Ross (accountant for Crosby).
Simmons, Bonnie (record librarian, KSAN).
Skinner, Rob (sales manager, KSAN).
St. James, Margo (guest, Dr. Hip Pocrates program, KSAN).
Stone, Alan (disc jockey, KMPX, KSAN).
Street, Dusty (engineer, KMPX; disc jockey, KSAN).
Sullivan, Jack (president, Metromedia Radio Division).
Tagnoli, Dawn (media buyer, Post, Keyes and Gardner).
Talbot, Bernadette (media buyer, Lennen and Newell).
Towle, Jack (salesman, KMPX).
Turner, Mary (promotion director, KSAN).

Voco, Abe Kesh (disc jockey, KMPX, KSAN).
Wachspress, How (chief engineer, KSAN).
Wehr, Don (owner and manager, Music City).
Weinberger, Stan (owner, Mr. Broadway).
Werfhorst, Marilee (general manager's secretary, KSAN).
Yochim, Bob (friend of Avery).
Young, Roland (disc jockey, KSAN).
Yurdin, Larry (media organizer).

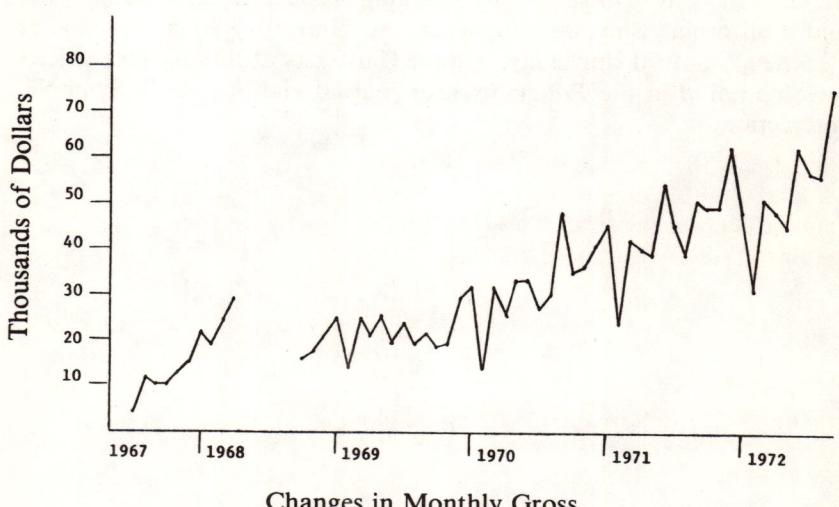

SUMMARY OF STATION SALES GROWTH 1967-72

Source: KMPX sales records; KSAN monthly sales reports.

ABOUT THE AUTHOR

Susan Krieger is a Visiting Assistant Professor in the Department of Speech Communication at the University of New Mexico. For a number of years she was involved in city planning agencies in both Connecticut and California. She has taught at the University of California at Berkeley, Stanford University, and the University of Illinois. Her articles have appeared in the *Policy Sciences* journal and Studies in Symbolic Interaction.